From Knowledge and Capacity Development to an Implementation Science: Policy Concepts and Operational Approaches

In Focus – Special Book Series

From Knowledge and Capacity Development to an Implementation Science: Policy Concepts and Operational Approaches

Editors

G. J. Alaerts and C. Zevenbergen

IWA PUBLISHING

Published by **IWA Publishing**
Unit 104–105, Export Building
1 Clove Crescent
London E14 2BA, UK
Telephone: +44 (0)20 7654 5500
Fax: +44 (0)20 7654 5555
Email: publications@iwap.co.uk
Web: www.iwapublishing.com

First published 2022

British Library Cataloguing in Publication Data
A CIP catalogue record for this book is available from the British Library

ISBN: 9781789063561 (paperback)

Contents

© 2022 The Authors From Knowledge and Capacity Development to an Implementation Science: doi: 10.2166/wp.2022.002

Editorial: From knowledge and capacity development to an implementation science: policy concepts and operational approaches

CONTEXT AND BACKGROUND OF THIS SPECIAL ISSUE

This Special Issue of *Water Policy* is dedicated to knowledge and capacity development (KCD). It offers selected contributions – analytical and review papers, and case studies – to the sixth International Symposium on Knowledge and Capacity Development held in May 2020[1]. This symposium was held 30 years after the first event in June 1991 which was convened by the IHE Delft and the UNDP (Alaerts *et al.*, 1991) to draw lessons from the lack of progress during the International Drinking Water Supply and Sanitation Decade 1981–1990 which aimed to 'provide water supply and sanitation to all by 1990'. The weak capacity at an individual level, organisational level and in the 'enabling environment' was identified as a fundamental impediment, and the symposium offered a strategy to address this. The 1990s and early 2000s also saw the emergence of initiatives to further conceptualise and explore KCD in public administration and governance frames. The sixth Symposium, thus, sits at a critical juncture: on the one hand, it draws from three decades of experience and, on the other hand, it looks forward to the middle of the century which is identified by many policy and research organisations as a significant date of concern regarding the state of, and the threats to the global water system (see also Alaerts & Kaspersma, this Volume). The now existing body of insight and experience makes a clear case for a more programmatic and professionalised approach to KCD and structured learning in sectoral organisations, national sectors and policies, as well as educational systems. Although the subject has gained much traction over the past decades, in practice many of these efforts have remained of a comparatively simple and discontinuous, project-based nature rather than being programmatic. With large challenges to the global water system arising in the next few decades – due to demographic and economic development, land use changes and climate change – knowledge and capacity stand to gain in prominence. With most countries now disposing of reasonably well-articulated policies, their effective implementation is the new challenge. Knowledge and capacity to underpin a new 'implementation science' are being called for.

Due to the emerging coronavirus pandemic, the sixth Symposium had to be converted into a virtual, global event straddling all time zones and featuring over 50 sessions with presentations, panel discussions, working groups and plenaries. The virtual format of this knowledge exchange and mutual learning platform did limit the opportunity for informal interaction and networking, but, on the other hand, allowed experts to participate who otherwise could not have spared the travel time or expense; it also lowered the carbon footprint to well below 1% of the physical convention. The conversion at short notice posed formidable logistic and substantive challenges to the organising team, and it forced speakers and participants to operate outside their comfort zones and interact solely through, sometimes, unreliable internet from their homes or offices in places such as provincial towns in Uganda, rural Utah, Kyoto and Khartoum. While, overall, the Symposium was very well

[1] The Symposium was organised by the IHE Delft Institute for Water Education, Delft, The Netherlands, with active support from the Asian Development Bank, the African Development Bank, the Organisation for Economic Co-operation and Development (OECD) and the National Water and Sewerage Corporation of Uganda.

attended and can be considered successful, the pandemic had a noticeable negative impact on the personal life of scientists and professionals – especially so in developing nations – who had to navigate prolonged restrictions and lock-downs and prioritise care for families and colleagues. This hampered contributors in their endeavours to prepare and submit papers for this Special Issue. Several prospective authors were unable to complete their submission and those who did, often needed much flexibility in deadlines, which explains the delay in the final publication.

PARADIGM SHIFTS IN WATER MANAGEMENT

KCD has developed over time, adapting to the evolving needs of society and the advancement of knowledge, tools and techniques. Increasing demands from a growing population, combined with the current large adaptation deficit of water infrastructure in the developing countries and the aging infrastructure in the richer countries, pose significant challenges to improve – or even sustain – our quality of life (Bhave *et al.*, 2016; Hall *et al.*, 2019; Wilby *et al.*, 2021). Comparing the present situation with that of the 1990s and early 2000s, fundamental shifts are apparent. First, global water management systems are becoming more complex and dynamic requiring a multi-scalar perspective, i.e., founded on a system approach connecting spatial and temporal scales. Second, society is expecting more social–ecological inclusiveness (inclusion of social and ecological aspects in decisions about water management); decisions in water management increasingly must engage the public and other stakeholders, including a representation of the ecological interests. Third, major advances in information technology and digitalisation are facilitating the generation and use of vast arrays of complex data in infrastructure and resource management processes. They allow better alignment of planning, design, engineering and asset management, in an iterative process that includes feedback loops (learning steps) leading to greater efficiency in infrastructure delivery. Also, finally, despite this forecasting capability, significant uncertainty remains regarding this future which means that any planning effort must avoid prematurely locking in future end-results through early investments and decisions; it must also operate in an iterative adaptive fashion, stepwise learning what works and what does not. Eventually, these shifts are helping the development of robust water infrastructure systems in the sense that they are more resilient and adaptive, and at the same time provide more flexibility. Yet, investments in water infrastructure are still lagging and their quality varies substantially. In most countries, the focus remains on managing current events and crises rather than on preventive actions (Hall *et al.*, 2019).

As competition for limited water resources is steadily growing, and as regulatory frameworks – on environmental impacts, land acquisition, financial sustainability, etc. — are expanding and becoming more complex, engineers, managers and policy-makers need to become more able to work across different disciplines, and to integrate water in other strategies and programmes such as land use, agriculture and environmental management. The rising constraints on water resources in river basins, and climate adaptation, necessitate more robust forecasting capabilities as well as cooperation across sectors and deeper engagement with civil society. Better knowledge is needed on how to enhance individual and organisational learning capabilities that will result in more effective capacity of individuals and in the transformational changes that must be made possible by improved organisational procedures.

To understand what the future may hold for KCD, the developments and trends in the water management domain must be identified which have the potential to impose fundamental changes on the current water management paradigm. Climate change is doubtlessly one of the most prominent changes, slowly panning out over the coming decades. The acceleration of the sea level rise and the changes in frequency and severity of extreme weather conditions indicate that climatic conditions are changing at an increasing rate (IPCC, 2022). Reliance on traditional incremental planning approaches based on extrapolation of past experience or on projections of future climate parameters must be questioned as they assume such projections are fully accurate, and that sufficient time remains to adapt. This holds particularly true for major infrastructure investments with long lead and life times and large economic and social impacts, and where today's decisions have long-term consequences locking in our future management (Haasnoot *et al.*, 2020). A fundamental change in the way we plan our infrastructure and management will be necessary, with a shift from incremental to transformative adaptation of water management systems; this will require a governance frame, new in ambition and in institutional setting (Fedele *et al.*, 2019; Hölscher & Frantzeskaki, 2020; Bloemen, 2021). As part of this new governance, decentralised systems will likely become more common. At present, still too much emphasis is placed on short-term solutions and incremental adaptation which constrains options to consider sustainable solutions for the long term (Watkiss *et al.*, 2020; Nicholls *et al.*, 2021). The alternative 'transformative' strategies to infrastructure planning are starting to emerge, involving back-casting, story lines and design-based approaches starting from an appealing, desired vision on the future (Shepherd *et al.*, 2018).

CONCLUSIONS AND RECOMMENDATIONS OF THE 2020 SYMPOSIUM

At the Symposium, the salient issues, conclusions and recommendations of the deliberations drawn by the participants are as follows.

- ***Instil a system perspective across all water-related research, training, planning and decision-making.*** Water issues are shaped by social and political dynamics and are wicked in nature, always involving trade-offs between competing goals and values. Assessing water issues in silos creates business cases for 'stupid' infrastructure that may yield short-term financial returns but may make societies more vulnerable in the long run. KCD activities offered by agencies, institutes and non-governmental organisations (NGOs), and which is focused on specific aspects of the water challenge, may risk compounding this problem, resulting in a patchwork of disconnected activities frequently lacking consideration for the systemic side-effects.
- ***Reduce capacity development activities which have a short-term or backward perspective.*** A significant share of current KCD in the water sector is based on scaling up 'best practices'. By definition, this approach draws on problems and solutions of the past, and may distract from creating new systems capable of adapting to new problems in the future. Also, when KCD providers engage in separate successive short-term pilot activities, they may learn from each attempt, but may also avoid responsibility for systemic consequences.
- ***Break the 'counting heads' mentality for measuring impact.*** KCD is still measured in terms of logistical indicators, such as numbers of people trained, years of schooling or regulations ratified. Research has conclusively shown that this does not capture actual changes in capacity. The sector needs to move beyond simplistic statistics, complement quantitative with qualitative methods and measure systemic impacts in ways that also speak to political decision-makers. Where feasible, more use should be made of randomised controlled trials.
- ***Improve inclusiveness and blended approaches.*** Achieving water-related sustainable development goal (SDG) involves action by societies as a whole, which requires trust and social acceptance for innovations and change. Focusing KCD on narrow groups of professional risks distracting attention from the stakeholders affected by new practices. Meaningful participation, in particular of women, youth and marginalised groups, is a long-standing good practice, but is still poorly applied in practice. KCD needs more diversity in blended approaches, to accommodate a wider range of audiences and learning styles.
- ***De-link KCD funding more often from confined project budgets.*** Too much of the KCD is linked to infrastructure projects, concentrating funding in one-off activities, not in areas of highest need. The 'last mile' of rural water supply, subsistence agriculture and the stewardship for essential aquatic ecosystems are particularly underserved. Requiring KCD providers to become financially self-sustaining, while sensible, does only benefit 'clients' who already have financial means. Societal capacity to achieve the SDGs includes public goods, and more mixed public–private financing models are needed to deliver on this.

OVERVIEW OF THIS SPECIAL ISSUE

The 12 papers in this special issue reflect recent work by scholars and professionals in their endeavour to understand how to effectively strengthen the capacity. They are drawn from the presentations at the Symposium and complement the above 'Conclusions and Recommendations'. While in the 1990s work on capacity mostly focused on understanding and exploring the very concept, and on 'training', in the early 2000s the initiatives appeared increasingly geared to more ambitious broad governance questions and the capacity of government administrations (*e.g.*, by the UNDP, the Asian Development Bank, the Canadian and British development agencies, the EuropeAid, the German GiZ, the OXFAM and other international NGOs). The work in the current special issue, however, is rather reflecting a shared concern to enhance the effectiveness of KCD interventions in general, or of existing programmes. For example, several papers assess the capacity of government and stakeholders to prepare for, and implement the complex water policies and programmes of the next decades. Other papers concern the development or piloting of didactic tools and frames that can be functional in the engagement with local communities and other stakeholders. Drawing from the six 'application arenas' proposed by Alaerts and Kaspersma (in this special issue), the papers in effect cover all these arenas albeit with a greater emphasis on arena 2, i.e., the development of complex skills and behaviours in particular in the context of engaging local communities and other stakeholders to prepare and implement more effective water programmes. These authors, however, also advocate urgent financial and political investment in broad-based capacity development to support the implementation of policy and institutional reform to counter these threats (arena 6). A second noticeable shift reflected in this special issue is that most authors hail from developing or middle-income countries suggesting that the subject is increasingly getting mainstreamed and receiving quality attention. The papers originate in Asia, Africa, Latin America, Europe and Australia. Finally, a sub-sectoral shift is also observed, from the formerly dominant role of water supply and sanitation to a broader coverage of river basin, delta and watershed management.

Taylor *et al.* propose a Stream Assessment Scoring System, a citizen-based science approach to strengthen local communities' insight into water quality, in order to engage them as stakeholders in monitoring and ultimately in decision-making tasks. Kale *et al.* developed a computer-based didactic tool that communities in Watershed Committees in water-scarce Rajasthan can use to better understand their groundwater aquifers and how these can be used in conjunction with surface water; this community-driven visual integrator notably offers a three-dimensional map rendering the 'invisible' groundwater visible and, thus, better manageable. Shibly *et al.* expand the applicability of the Motivation and Ability Framework to assess the social adaptability of technical plans for managing coastal regions in Bangladesh. The critical importance of a more appropriate use of adult learning theory and approaches is highlighted by Russell and Ottow who analyse the impact of three learning events and advocate action-oriented learning and better alignment of the delivery system with the learning objectives. Del Pozo *et al.* review training for climate-related services and also here, conclude that such training suffers from being too much one-size-fits-all, recommending the wider use of more locally relevant educational frames and case studies, as well as of up-to-date educational and pedagogic principles.

Martel *et al.* employ a 'governmentality' framework to analyse the capacity development processes of three river rehabilitation projects in South Africa. The analysis reveals that the three projects have different governmentalities; although this produces different capacity development modalities, they identify common elements that support capacity development such as actionable information, a conducive context for learning, and the role of intermediaries. A wide (though preliminary) assessment is offered by Kabir *et al.* of the capacities that will be required in government, communities and education to implement Bangladesh's ambitious Delta Plan 2100. Importantly, the study identifies weaknesses and tasks for the specialisation of individual professionals, for managerial change in the institutions, and for the higher education system. In a similar vein, Sabbatini and Indij study current practices and the modalities on different continents for scaling up and accelerate capacity development which would be necessary to attain by 2030 the SDGs for integrated water resources management and transboundary water management. They notably call attention to the gap between the broad recognition of the general Agenda 2030 goals and the (lack of) concrete specific understanding of, and action for the more technical water objectives. All papers concur that the availability of locally pertinent data and information is crucial to support the development of local decision-making capacity and arrive at solutions that are truly compatible with the local conditions. Conallin *et al.* present an approach for Capacity Building through Research drawing on five pilot case studies. They assess the appropriateness of the approach using the Dutch Strategy Evaluation Protocol, and show it leads to better aligned long-term strategies and the advice to, i.e., develop cross-disciplinary networks to enhance the capacity of local water management institutions.

Pillai and Narayanan assess CANALPY, a transdisciplinary initiative for knowledge co-production to address the capacity gaps in tackling the sanitation problem in a small town in Kerala, India, particularly to rejuvenate its canals. Researchers involved in the initiative worked with local volunteers, citizens, government officials and the urban local body for the production and sharing of knowledge to generate context-specific interventions. They explore the knowledge co-production process and the various enabling drivers and barriers to the process, stressing the two-way communication and engagement between government and stakeholders – local communities but also the other water-related sectors. They notably highlight the major challenge that the public-sector institutions (water and city agencies) face to operate effectively outside the narrow confines of their traditional organisational protocols. Dektar *et al.* assess the capacity weaknesses of the private water operators which provide piped water supply in rural Uganda, and they find that the operators lack capacity both on technical matters (such as on solar-powered pumps) and on customer relations, which combine to depress the customers' willingness-to-pay. Finally, Alaerts and Kaspersma report the case of the national-scale implementation of Indonesia's policy reform of its irrigation sector over a prolonged time; they describe how a progressive iterative capacity development process simultaneously strengthened national and local governments, and Water User Associations, while at the same time ensuring that lessons learned from field application were used to enhance the national and local regulatory and administrative structures.

The papers share key concepts and approaches leading to the following overall conclusions: (1) the role of local communities and committees in water resource management is receiving ever more recognition, and studies and pilots are exploring new tools, often internet-based, to assist in developing local knowledge and capacity to enhance the decision-making on water management at community level, and, thus, make these more resilient; (2) training and capacity development remain too often conventional, and more effort needs to be invested in rendering them locally-specific using culturally adjusted frames, locally relevant case studies, and up-to-date pedagogical approaches; it is critical that KCD become systemic and a long-term engagement with outcomes and impact specified in broad terms; (3) institutional capacity is what matters most, at the level of government, water organisation and community; (4) knowledge and capacity will increasingly guide our responses to counter the threats to our water system; they need to be substantially enhanced to support the emerging implementation science to turn our ambitions and policies into effective action and investments; and (5) investing in capacity is economically sound, also in the short run, and brings significant returns both to those benefiting from KCD (thus creating a powerful incentive) and to the state.

For the future we would call for more scientifically sound treatises, based on robust statistical analyses of impacts, randomised controlled trials or on rigorous comparison of with/without and before/after situations. Such approaches generally would necessitate much higher funding and a more coherent and longer time frame. Currently, KCD activities are still too often set up and financed for short-term and pragmatic purposes with the expectation that well-described capacity solutions get developed, not studied. This is, therefore, an oxymoron in a sense: KCD is about setting up learning processes in institutions and in society, but in today's practical application, KCD activities are mostly financed to only deliver a product specified in advance, not to arrange a longer time frame and process to structurally learn from these activities and discover sustainable development paths forward.

DATA AVAILABILITY STATEMENT

All relevant data are included in the paper or its Supplementary Information.

Guest Editors
G. J. Alaerts
IHE Delft Institute for Water Education, Delft, The Netherlands

C. Zevenbergen
IHE Delft Institute for Water Education, Delft, The Netherlands

REFERENCES

Alaerts, G. J., Blair, T. L. & Hartvelt, F. J. (1991). *A Strategy for Water Sector Capacity Building.* Int. Institute Hydr. Envir. Engrg./United Nations Development Programme, Delft/New York.

Bhave, A. G., Conway, D., Dessai, S. & Stainforth, D. A. (2016). Barriers and opportunities for robust decision making approaches to support climate change adaptation in the developing world. *Climate Risk Management 14*, 1–10.

Bloemen, P. (2021). *Views on Climate Change.* Analytical concept, Memorandum [2021 04 15].

Fedele, G., Donatti, C. I., Harvey, C. A., Hannah, L. & Hole, D. G. (2019). Transformative adaptation to climate change for sustainable social-ecological systems. *Environmental Science & Policy 101*, 116–125. https://doi.org/10.1016/j.envsci.2019.07.001.

Haasnoot, M., Kwadijk, J., Alphen, J., Le Bars, D., Hurk, B., Diermanse, F., Van der Spek, A., Oude Essink, G., Delsman, J. & Mens, M. (2020). Adaptation to uncertain sea-level rise; how uncertainty in Antarctic mass-loss impacts the coastal adaptation strategy of the Netherlands. *Environmental Research Letters 15.* doi:10.1088/1748-9326/ab666c.

Hall, J. W., Aerts, J. C. J. H., Ayyub, B. M., Hallegatte, S., Harvey, M., Hu, X., Koks, E. E., Lee, C., Liao, X., Mullan, M., Pant, R., Paszkowski, A., Rozenberg, J., Sheng, F., Stenek, V., Thacker, S., Väänänen, E., Vallejo, L., Veldkamp, T. I. E., van Vliet, M., Wada, Y., Ward, P., Watkins, G. & Zorn, C. (2019). *Adaptation of Infrastructure Systems: Background Paper for the Global Commission on Adaptation.* Environmental Change Institute, University of Oxford, Oxford.

Hölscher, K. & Frantzeskaki, N. (2020). *Transformative Climate Governance. A Capacities Perspective to Systematize, Evaluate and Guide Climate Action.* Palgrave Macmillan, London.

IPCC (2022). In *Climate Change 2022: Impacts, Adaptation, and Vulnerability. Contribution of Working Group II to the Sixth Assessment Report of the Intergovernmental Panel on Climate Change.* Pörtner, H.-O., Roberts, D. C., Tignor, M., Poloczanska, E. S., Mintenbeck, K., Alegría, A., Craig, M., Langsdorf, S., Löschke, S., Möller, V., Okem, A. & Rama, B. (eds.). Cambridge University Press, Cambridge/New York.

Nicholls, R. J., Hanson, S. E., Lowe, J. A., Slangen, A., Wahl, T., Hinkel, J. & Long, A. J. (2021). Integrating new sea-level scenarios into coastal risk and adaptation assessments: an ongoing process. *WIREs Clim Change 12*, e706.

Shepherd, T. G., Boyd, E., Calel, R. A., Chapman, S. C., Dessai, S., Dima-West, I. M., Fowler, H. J., James, R., Maraun, D., Martius, O. & Senior, C. A. (2018). Storylines: an alternative approach to representing uncertainty in physical aspects of climate change. *Climatic Change 151*, 555–571.

Watkiss, P., Wilby, R. L. & Rodgers, C. A. (2020). *Principles of Climate Risk Management for Climate Proofing Projects.* Asian Development Bank, Manila.

Wilby, R. F., Lu, X. F., Watkiss, P. & Rodgers, C. A. (2021). Towards pragmatism in climate risk analysis and adaptation. *Water Policy 23*(S1), 10–30. https://doi.org/10.2166/wp.2021.232.

© 2022 The Authors From Knowledge and Capacity Development to an Implementation Science: doi: 10.2166/wp.2022.301

Facing global transitions in water management: Advances in knowledge and capacity development and towards adaptive approaches

G. J. Alaerts [a,*] and J. M. Kaspersma [b]

[a] IHE Delft Institute for Water Education, Delft, The Netherlands
[b] Department of Flood Risk Management, Deltares, Delft, The Netherlands
*Corresponding author. E-mail: g.alaerts@un-ihe.org

GJA, 0000-0002-9537-2928

ABSTRACT

The significance, approaches, and instruments of knowledge and capacity development (KCD) in water management are reviewed, and priorities for the future are proposed. These concepts have become more mainstream, critical in helping water organisations decide on and implement policies, and generating economic returns. Their application still tends to be often limited to education or 'training'. KCD requires an understanding of the physical world, how institutions must be strengthened to manage it, and how pedagogical and knowledge-management tools, in turn, strengthen the institutions. The private sector first applied knowledge management. The international development theory highlighted the deficiency of governments in implementation capacity. The health and environmental communities are developing an 'implementation science' to enhance the capacity to operationalise know-how faster. Advances in KCD include the following: (1) knowledge and capacity converging in nested levels (individual, organisational, institutional, and societal) to cause effective action; (2) six arenas/contexts of KCD application; and (3) pedagogy and knowledge-management through which learning occurs and knowledge is imparted. KCD is a 'sticky', slow process. Policy analyses tend to overlook the role of KCD. The water sector is facing acute challenges: a new one of building resilient water-and-land systems and adapting to climate change, and the outstanding one of achieving Sustainable Development Goals (SDGs). Thus, the current KCD must be scaled up, and also structured with a longer-term perspective to support change and reform at policy and organisational levels building on iterative adaption. Policies should become pro-active shaped by modelled forecasts; and organisations more able to change and adapt to future scenarios that are complex, uncertain, and evolving rapidly. Enhancing the capacity to implement policies, establish 'learning organisations', and design iterative adaptive pathways requires sustained political commitment. While adopting a long-term programme, supportive KCD activities should stay realistic and manageable.

Key words: Adaptation pathways, Capacity development, Economic return, Institutional development, Knowledge management, Learning

HIGHLIGHTS

- KCD improves water management.
- Instruments are drawn from pedagogy, education, and management.
- Capacity to implement policies is often deficient.
- Water sector must prepare for transitions due to economic growth and climate change, demanding capacity for adaptive management.
- Looking forward, a new agenda is proposed to enhance implementation capacity and design iterative adaptive pathways.

INTRODUCTION

Water management is carried out by stakeholders and institutions – the organisations and the rules that govern and administer society's activities. They function by virtue of their capacity to make decisions and get these implemented, supported by a combination of their knowledge and (political) commitment (North, 1990; Sveiby, 2001). Knowledge is necessary to articulate the subject matter – 'what we need to do or achieve' – as well as the implementation process – 'how we will do it'. Knowledge is central to capacity which is the capability to decide on goals, act to attain these, and to learn from this experience. This paper aims to review the advances made in the past three decades in practising knowledge and capacity development (KCD) by assessing the complementary roles of institutions, knowledge, and policy design and implementation. It will argue that new systemic challenges are arising in the next three decades that will emphasise the need for more effective policy implementation for adaptive water management. This forward-looking review thus sits at the mid-point of a period spanning about six decades bounded by the first identification in 1991 of weak institutional capacity as one main reason for the partial failure of the 1981–1990 UN Drinking Water Supply and Sanitation Decade (IDWSSD) (Alaerts *et al.*, 1991; O'Rourke, 1992) and by about the year 2050 that according to many water and climate forecasts risks becoming a tipping point and by which time better resilience should be achieved (see further). While the paper concerns primarily countries in development, the knowledge and capacity challenges are common to all countries irrespective of their wealth.

Epistemology has defined knowledge as the capacity to act (Polanyi, 1958). Knowledge is acquired by an individual through reasoning, discovery, and learning. Whereas in industrialised economies and in firms the literature is concerned with *knowledge management* to enhance performance, in public management in developing economies these processes are generally considered *capacity development* (Alaerts, 1999). In economies with a robust education system, strong governance and high levels of communication and transparency, new knowledge can dissipate relatively fast across institutions and society and be applied through decisions in commercial transactions, regulations and procedures decided at national and local government levels. Citizens and their representatives participate in these processes. In developing economies, on the other hand, new knowledge travels more slowly through a nation. Policy innovation and reform that imply redistribution of power and delegated decision-making can only be implemented stage-wise as the prospective new actors need time to acquire the capacity to understand their new role, agree to it, and start acting accordingly. Policy reform thus rests on simultaneity of political decision and capacity. For capacity to be acquired, political will must exist, and to inform sound political decision-making, capacity is a prerequisite. The effective impact of a policy is predicated on implementation capacity.

KCD FOR INSTITUTIONAL PERFORMANCE

Knowledge and capacity

Knowledge is central to institutional capacity yet it rarely receives explicit strategic attention. Typically, it is identified as an issue when its absence is considered the cause of current poor performance, but is rarely subject of strategic policy and sector-wide management. Alaerts & Kaspersma (2009) propose an analytical framework that emphasises that capacity and knowledge must exist at the following four nested levels: (1) *individuals* need to possess generic and functional knowledge in the meanings of factual knowledge (in the sense of the scientific consensus on discoverable and refutable physical, economic and behavioural phenomena), understanding, skills, and attitudes; (2) *organisations*, i.e. institutions in which the individuals work or interact must dispose of specific problem-solving heuristics ('operational practices') by aggregating individual staff knowledge but also by rendering this knowledge effective and acquire and embed new knowledge in these practices – the organisations becoming 'learning organisations'; (3) the institutional *enabling environment* with the institutions of government, policies, regulations, and agreements in which these organisations operate; and (4) *civil society*, i.e. civic groups or society, as they are seeking to understand their challenges, negotiate shared values, and acquire new knowledge.

For conceptual and operational reasons, KCD concerns itself with learning and knowledge generation and sharing. Many insights are drawn from traditional educational and pedagogical principles but others were formulated only in the second half of the 20th century, often in the context of management or behavioural sciences. Epistemology highlights the difference between explicit knowledge that is formalised, e.g. in text books and can be easily transferred, and implicit (tacit) knowledge that cannot be codified and needs to be discovered through example and trial and error, like riding a bike (Polanyi, 1958). Generally, knowledge acquisition presupposes a willingness to discover new insight, or learn, and takes five main routes. Two routes occur within an individual's own cognitive capability through deduction or induction. Three other routes are exogenous (or a hybrid): experiential learning (e.g. as learning-by-doing [e.g. Kolb, 1984] or structured experiential learning [Pritchett *et al.*, 2013]), absorption of external knowledge (e.g. in classroom teaching and most training) and mimicking (e.g. in formal or informal mentoring and peer-learning in organisations, communities-of-practice or networks [e.g. Schein, 1978; Guilmette, 2009; Mangin & Dunsmore, 2014]). Learning processes in which we are emotionally engaged or discover by experiment generate knowledge that sticks better to our memory (e.g. Niedenthal *et al.*, 2006). Finally, pedagogy and behavioural sciences offer insight into methodologies for effective teaching and knowledge transfer in both physical teaching and internet-based education.

Knowledge in institutions

The insight that knowledge is also an institutional attribute emerged in the mid-20th century when knowledge as a productive economic input became valued in macro-economic models (e.g. OECD, 1996). Knowledge operates by inducing changes in behaviour and decision-making and adds value to society and firms, creating a competitive advantage by taking the form of an educated workforce and of know-how and innovation capability in the organisation (as in intellectual capital and patents). When new policies are to be implemented in complex environments where many institutions are actors – typical for water management – the capacity of all institutions often needs to be developed. Institutional capacity can be developed through, e.g. formal education (for the longer term); training for imparting competencies, skills, and attitudes; human resources management; research and innovation; development of policy and administrative arrangements based on, and reinforcing knowledge; peer-learning inside and among institutions; coaching and mentoring; and learning-by-doing. A central element in institutional learning is structured monitoring and evaluation (M&E) of results against planned outcomes and feeding back lessons learned during this implementation. Public institutions are still slow in adopting learning strategies though the corporate sector already in the 1980s and 1990s started to treat knowledge as a firm's asset that needs to be managed for the organisations to be effective. Firms like Toyota, General Electric, and Asea Brown Boveri pioneered new management approaches. Senge (1990), Argyris (1992), Nonaka & Takeuchi (1995), Sveiby (1997) and Easterby-Smith & Lyles (2003), among others, developed the discipline of organisational knowledge management. It applies concepts of human resources management and competences management; the articulation and retaining of experience, such as in best-practices and knowledge repositories; and internal learning protocols as well as managerial procedures that create space for experimentation including occasional failure[1]. The recognition that knowledge is a fundamental attribute of public sector policy and

---

[1] The World Bank, a leading development agency, for example, adopted new internal procedures in 1999 to foster knowledge creation, dissemination, and retainment. It started to require that each study or loan proposal be formally reviewed at least twice by three independent peers in open meetings; best-practices from field projects to be rigorously documented and shared across regions through communities of practice; and human resources procedures to attract these skills and reward such staff behaviour. It set out a strategy and programmes for

administration is gaining traction only slowly. Here, it leads to more effective policies and improved problem-solving heuristics in administration and policy implementation. Some countries, such as The Netherlands, have explicit knowledge-management policies for 'top' sectors including that of water (Ministry of Economic Affairs, Agriculture & Innovation, 2011; CBS, 2012).

Definitions

Capturing KCD in definitions has proved challenging; they tend to reflect two orientations. Morgan (1993), whose work inspired many development agencies, casts it as a fundamental governance endeavour: 'Capacity development is the ability of individuals, groups, institutions and organisations to identify and solve development problems over time'. This broad perspective at the same time also is a weakness as it suggests capacity development is almost synonymous with development itself, without providing insight in how to operationalise ability. Other authors stress both the role in self-determination and the instrumental nature of KCD, e.g. 'KCD is the capability of a society or community to identify and understand their development issues, to act to address these, and to learn from experience and accumulate knowledge for the future' (Alaerts, 2009).

KCD IN POLICY DEVELOPMENT AND IMPLEMENTATION

KCD as operational practice

Over the past three decades, KCD has tended to take place in approaches that can be pragmatically grouped into six arenas characterised by purpose, context, processes, and instruments. Some of these arenas are distinct, but others overlap or intersect, or are nested in others. They are as follows: (1) development of skills for well-defined executive tasks; (2) development of complex behavioural skills; (3) education; (4) research, innovation, and R&D; (5) hardware and software for communication and decision-making; and (6) policy and administrative reform.

Arena 1 activities concern short-term development of vocational, cognitive, and sometimes interpersonal skills and competencies needed to carry out specific straightforward tasks of generally technical or mono-disciplinary nature, such as pipe-laying, water-meter reading and repair, accounting, community organising, mathematical modelling, etc. These activities have well-defined objectives, draw upon established training approaches (notably classroom training, workshops and coaching) and yield easily measurable outcomes with a clear causal training–outcome relation.[2] These activities are intuitively worthwhile and uncontroversial.

Arena 2 activities encompass development of more complex competencies, often of managerial or behavioural nature. This more open-ended KCD may concern professionals in managerial or other leading positions, but also diverse groups in civil society which need to be coached to identify and address their local challenges. It typically necessitates a broader more specialist toolset of teaching, structured and experiential learning, and personal discovery in a coaching frame; much of the knowledge imparting occurs through self-assessment and simulations. It draws on approaches and instruments of Arena 1 but also builds on networks, communities-of-practice, and peer-learning and peer-assessments. In local communities, processes are applied such as Participatory Rural Assessment, actor analysis and behaviour models, and Motivation and Ability frameworks. Aims and scope vary. Minimally they include developing personal management and leadership skills of individuals (on technical tasks, organisation, leadership, results-orientation, etc.) to enhance staff and institutional performance. In a

knowledge-sharing across the world and launched its Global Delivery Initiative that invests heavily in documenting experiences and making them available across the world through networks, repositories and operations (World Bank, 1998, 2018).

[2] In certain countries strong master–pupil relationships have fostered traditional forms of training in trades and crafts such as the *Meister* programmes in Germany and *shokunin* in Japan. These combine mimicking and structured learning in a coaching frame, such as in internships.

broader scope, they form part of Arena 6 in transitions of (part of) organisations by enhancing the skills sets of individual staff to support the redesigned organisation.

Arena 3 concerns education and knowledge-generation. Commonly these aim to prepare young generations for their future and help develop their basic knowledge, cognitive, vocational and interpersonal skills, and attitudes, as well as their capability for continuous further learning. A vast educational infrastructure exists to achieve these aims and simultaneously set up the platform for research (Arena 4). The KCD instruments range from classroom teaching to experiential learning, coaching, peer-learning, mimicking and mid-career education; pedagogy offers menus of tested instruments and approaches for different contexts and purposes. The outcome of education is generally well discernible at aggregate level in the form of the individual's and economic development, although the extent of the contribution of schooling to personal development is a matter of debate. The 70–20–10 rule suggests that – at least in a professional context – executives acquire 70% of their knowledge from hands-on experience, 20% through social learning, mentoring, collaborative learning and other interaction with peers, and only 10% from formal traditional courseware instruction (Lombardo & Eichinger, 1996). In a world of progressing complexity, advanced specialised knowledge has premium value yet many problems are of a wicked nature, i.e. are difficult to describe precisely in advance, subject to changing circumstances and of transdisciplinary nature. This calls for generalist knowledge. Some professionals and institutions may be able to accommodate both specialist and generalist knowledge; Uhlenbrook & de Jong (2012) argue that professionals are best equipped with T-shaped competencies. In most cases, however, effective heuristics will increasingly depend on the capability of individuals and institutions to collaborate in teams or networks able to combine different specialised competencies.

Arena 4 focuses on the discovery and generation of knowledge, such as in scientific research, R&D, innovation, and citizen-based science. The processes involved often are partially integrated with those in the other arenas. Arena 5 concerns the development and use of hardware and software upon which much of the communication, information sharing, training, simulations and decision-making is resting nowadays. The deeper penetration of hand-held phones and internet-based tools keep contributing to faster and wider knowledge dissemination and decision-making.

Finally, in Arena 6 the KCD ambition extends to the support to a change, reform programme or transition in organisations, policies or even assemblies of sectoral actors and institutions such as ministries, (water) agencies as well as political entities. In many instances – as water is affected by the behaviour and decisions of households and other users – this may also encompass civic stakeholders; e.g. farmers play key roles in irrigation, and industry in water pollution control. Focus lies on both policy articulation and its implementation. Like in Arena 2, KCD is rarely of simple technical nature and generally implies articulation of preferences and political choices for which consensus needs to be sought and sustained over prolonged periods, often 10–20 years. Such organisational change programmes are common in the corporate world but less so in the public sector or academic institutions, and when they exist, are often not well documented. Examples for the water sector refer to, e.g. the Uganda National Water and Sewerage Corp. and other African utilities (Mugisha *et al.*, 2007; Mugisha & Brown, 2010), the Philippines and Malaysia (Jensen, 2019) and the Netherlands Ministry of Transport and Water Management (Metze, 2008). Recent studies propose 'social learning' to enhance management, e.g. for irrigation in Australia (Nikkels *et al.*, 2019). Depending on scope, ambition and readiness, the duration of a KCD intervention likely takes many years if not decades. KCD outcomes are not well quantifiable or predictable and impact may have to be assessed through proxies that measure the institution's effectiveness on Key Performance Indicators or through perception surveys. A direct causality between interventions and indicators is often tenuous. Critically, in Arena 6, KCD intervenes in the management purpose, style and internal workings of the organisation, and can succeed only if political commitment exists to adopt and sustain the changes that re-arrange power. Though many sectoral assessments identify organisational weaknesses and the 'need to change' to achieve more effectiveness, political, institutional and often personal reluctance create roadblocks. Beside the complexity of the problem analysis and the KCD interventions at such large scale, often across several institutional jurisdictions, and the political contentiousness, the ultimate goals, pathways, benefits and risks of the reform process are generally not well known up-front and need to be discovered during step-wise adaptive implementation. Activities typically include approaches and instruments of Arenas 1 and 2 aimed at specific groups of stakeholders and decision-makers, beside more process-oriented tools such as Results Frameworks to monitor and evaluate interim achievements, identify and address constraints, reveal unanticipated changes in the environment or internal objectives, and adapt to these new circumstances (see The next challenge: KCD to support adaptive policy implementation).

Policy implementation as bottleneck

It remains challenging for notably developing nations to design and implement policies that are effective and help meet the targets of, e.g. the Sustainable Development Goals (SDGs). The High-Level Forums on Aid Effectiveness (Paris, 2005; Accra, 2008) identified weak institutional capacities for policymaking and implementation as the key impediment (OECD, 2019a). Reviewing international development performance, Andrews *et al.* (2017) conclude that implementation of policy is the main challenge, more than its design. Implementation capacity is at risk also in developed economies; notwithstanding tall ambitions to boost infrastructure and adopt green pathways, many initiatives are slow to materialise and generate realistic investment opportunities (OECD, 2021; Alaerts, 2022). The 2009 American Recovery and Reinvestment Act to boost infrastructure spending after the 2008 financial crisis partly failed due to a shortage of 'shovel-ready' proposals. In the Netherlands, concern is growing that government is increasingly beholden to short-term scoring with policy and legislation, at the expense of implementation realism (NRC, 2020). The articulation and implementation of realistic policies and investment programmes concerning water are especially challenging compared to, e.g. energy and transport, because of their complexity (as discussed below).

The process of articulating and implementing effective policies has been studied extensively. Kuhn (1962) posits paradigm shifts as the basis for scientific (and by extension, political) revolutions driven by growing consensus among experts regarding progressing dysfunction of the prevailing paradigm. Van der Brugge *et al.* (2005) show that institutions and policy domains tend to stability (with no or minimal fundamental change), i.e. internal optimisation rather than system innovation, to minimise transaction costs (North, 1990) or because policies are captured by groups of actors seeking the status quo (Meijerink & Huitema, 2009). According to the Multiple Streams Theory, policy changes occur when three streams (of problems, policies, and politics) converge (Kingdon, 1995). To achieve effective and efficient policy setting, impact monitoring and policy implementation must be approached as knowledge-intensive. The Advocacy Coalition Approach (Sabatier & Jenkins-Smith, 1993) stresses the crucial role of knowledge and longer-term learning processes in the shaping of the coalitions. The Contextual Interaction Theory (Bressers, 2007) identifies motivation, information, and power as decisive factors for change. To these three change factors and to the three 'streams', before, one should add the role of personalities (personal ambitions of decision-makers) as well as institutional inertia that may work against change. Gidron & Bar-Almog (2010) recommend to observe policy change over a period of a decade or longer to find out how policy analysis shapes the agenda and how learning takes place. Also, Rotmans *et al.* (2001) emphasise learning as central to policy transitions. Most literature on policy development is geared to industrialised economies which possess a good governance architecture and educated citizenry. The Contextual Interaction Theory therefore highlights the limitations of replication of institutional models and their intrinsic knowledge from one cultural and political context to another (Bressers, 2007; Bressers & de Boer, 2013).

Policy change is particularly problematic where the government is at the same time a main actor in its implementation, because public sector agencies are more subject to rigidities and inefficiencies than the private sector. This challenge notably affects developing countries where bureaucracies tend to be more prone to inertia and elite capture, and where knowledge on implementation is less easily available.

Also, health services and environmental management now consider weak implementation capacity a systemic challenge. However, the focus here is on more rapid and systematic uptake of research results into practice (for the health sector 'from lab bench to bedside', down from the average 17 years), taking into account the institutional context and applying a structured monitoring framework. They appear less concerned with the management of the environmental or health management system itself (Bauer *et al.*, 2015; Hering, 2018)[3].

KCD for reform

While the above suggests that a deeper understanding of policy and institutional reform and the role of KCD has evolved over the past decades, a discrepancy is often observed between the diagnostics – the analysis of why a system is functioning poorly – and what improvements can be achieved by KCD on the ground. Examples of diagnostic frameworks are those of EuropeAid (2005), UNDP (2007), ECDPM (Keijzer *et al.*, 2011), ADB (2011) and GiZ (2015). In line with Morgan's definition (see higher), these diagnostic frameworks generally have ambitious analytical scopes covering a sector, or a set of policies and organisations regarding their policy and governance (Arena 6) but are less clear on what actual KCD programme would be able to develop this capacity. They tend to fall back on common intuitive instruments, such as education and training of individual staff, that can be executed in a project's timeframe of 3–4 years. Possible causes for the discrepancy are as follows: (i) over-reliance on exogenous inputs instead of building on local initiatives, (ii) underestimation of the need of political negotiation (e.g. Meijerink & Huitema, 2009) to secure commitment, even on part of the goals, (iii) the assumption that a change process can be planned rather than built up iteratively, and (iv) the high cost and duration of the engagement.

NEW TRANSITIONS IN THE WATER SYSTEM AND MANAGEMENT

From reactive to pro-active planning

Current global policy narratives on water are shaped by the SDG6 'to achieve universal access to safe drinking water and sanitation for all by 2030', halve pollution and apply Integrated Water Resources Management. It continues the 1981–1990 IDWSSD and the Millennium Development Goal 7 (2000–2015). These goals and their implementation strategies were based on scaling up past experiences, extrapolating historical data without addressing the growing implicit competition for water caused by the simultaneous goals to increase food and energy production and job creation in a world where water is already scarce. More recent policy discourses, on the other hand, are founded increasingly on forecasts and scenarios derived from mathematical simulation models based on deeper insight in these interconnected systems. The suites of climate change models for instance describe relatively reliable future scenarios linking changes in temperature, sea level and precipitation and evapotranspiration with locations. With expanding knowledge of these systems and more powerful simulation capability, planning will become more able to forecast non-linearity, competition and discontinuity in the water system.

---

[3] Two dedicated journals exist for the health sector: *Implementation Science* (Springer) and *Implementation Research and Practice* (SAGE Journals).

Transitions in the water system

Water sits at the nexus of several natural and social systems that require water to thrive (health, food production, ecosystems, etc.) and, in turn, decide on policies that impact heavily on water. Demographic and economic growth are the main stressors on water. Water over-abstraction and land use changes, including urbanisation, increasingly determine local water excess or shortfall, and quality. Most rivers and lakes are already heavily over-used. In many places the abstraction rate from aquifers exceeds recharge; concern is growing across the globe – from the Ogallala aquifer in the mid-west US, to south-western Europe, India's Rajasthan and China's north-eastern Hebei-Shandong region – that many local stocks will run dry in the next 2–3 decades unless drastic measures are taken to return to sustainable abstraction rates (Famiglietti, 2014; Turner *et al.*, 2019; Gleeson *et al.*, 2020). Over-pumping in urbanised areas, or for irrigation and land reclamation, is now causing land subsidence, by 10–30 cm per year, over vast areas in Indonesia, the Beijing-Shandong basin, Mexico City, Central Valley (US), the Ganges and Indus plains, the Teheran plain, to name a few. Much of this subsidence started only in the 1970–1980s. Herrera-García *et al.* (2021) estimate that from 2010 to 2040 this will expose a population of 480–660 million people and 12–16% of global GDP to intensified flooding. This starkly illustrates how compartmentalised water exploitation and poor land management are exhausting water resources while at the same time dramatically increasing flood risk. In the mid-term, the main drivers for aridification and increased flooding are conversions in land use that channel rain run-off faster to the river, such as by removal of forest canopies and marshland, and urban 'hard surface' such as roofs, roads and parking lots. At continental scale, vegetative cover evaporates moisture into the atmosphere that feeds precipitation elsewhere. In South America, the Río de la Plata basin depends on evaporation from the Amazon forest for 70% of its water resources. Thus, intensive conversion of forest to urban and agricultural land will have a major negative impact on water availability (Van der Ent *et al.*, 2010).

Water, ecosystems, and biodiversity exist in critical mutual dependence. On an estimated 23% of the global terrestrial area, current biomass productivity is estimated to have been depressed by land and water use changes and other interventions, compared to the undisturbed situation; this is affecting 36% of all cropland, pasture and forestry systems and 15% of natural areas. Biodiversity and ecosystems are by themselves very productive in a narrow economic sense (fish, timber, and produce) but the combination of landscape and vegetation ('green infrastructure') in turn ensures a healthy regenerative hydrologic system. OECD (2017, 2019b) estimates the current global productive value of ecosystems and biodiversity at about US$140 trillion, or 1.5 times global GDP, yet these systems are being degraded at an increasing rate. Biodiversity loss was estimated at 34% by 2010 and is projected to continue with about 10% of additional loss up to 2050, a juncture that may turn critical with respect to reversibility in land and soil degradation jeopardising water security (Van der Esch *et al.*, 2017; WWF, 2020). Studies have assessed the overall implications of combined water (over-)use and land use change. Drought appears as the deadliest physical hazard with at present an estimated 3.6 billion people living in areas that are water-scarce for at least one month per year, to increase to 4.8–5.7 billion by 2050 (i.e. 55–65% of the world population) with over 40% of the world population living in water-scarce river basins under severe water stress, up from about 16% in 2010 (OECD, 2012; Sadoff *et al.*, 2015; WWAP, 2018). Thus, the world started transitioning from a place that is predominantly wet to one that is predominantly arid. Driven by this increasingly unsustainable dynamic, the World Bank (2016) forecasts a likely decline in several countries' GDP of up to 6% by 2050 caused by water-induced losses in agriculture, health, income and property, with some regions in the world facing sustained negative growth.

Climate change is projected to significantly increase the population facing water scarcity and/or major river floods in the 21st century. The time frames for the scenario forecasts extend to 2050, 2100, and beyond (IPCC,

2021, 2022). Schewe *et al.* (2014) calculate that for each degree of global warming, an additional 7% of the global population will be exposed to a decrease in renewable water resources of at least 20%, and on average, 4% of the global land area will see, for each degree of global warming, a decrease in renewable groundwater resources of more than 30%, and 1% by a decrease of more than 70% (Portmann, 2013). Thus, by the mid-century most regions will be deeply affected by such changes.

The above highlights, first, that policy design is transitioning from a paradigm determined primarily by the extrapolation of historical data sets, to one driven by simulated scenario forecasts with a focus on the situation by about 2050. Second, the year 2050 may represent a tipping point with the water sector looking fundamentally different in the second half of the century than in the first. And third, where the conventional paradigm assumed that all demand – by households, agriculture, industry, etc – could be satisfied and that any negative impacts of water appropriation could be mitigated through add-on measures, the new paradigm accepts that the water system has inherent deep constraints and seeks increased efficiency of water use, protection of land and ecological systems, reduction of the 'water footprint' of economic activity, and more reliance on a circular economy including efficient water allocation. Capable and well-informed institutions will be essential to all policy and planning, and their implementation.

Transitions in sector management

Water management is also undergoing transformations at institutional level. In the corporate sector complexity and uncertainty are being recognised as new systemic features (McGrath, 2011). For the water sector such transitions are reflected notably in legal and regulatory frames, policy documents and social analyses, which converge in the operational tasks, priorities and goals as perceived by the leadership of water utilities and agencies. These transformations are not well analysed yet. However, the authors conducted an exploratory survey and literature review for the situation of the regional (drinking water) utilities and the regional wastewater and flood management agencies (Water Agencies) of The Netherlands spanning the period 1970–2030[4]. Though confined in scope and methodology the outcomes are considered meaningful and also reflecting trends in higher-income economies as this country is considered to be performing well on governance and overall capability (OECD, 2014). Table 1 provides a qualitative synthesis of the declared perceptions. The leadership is faced notably by expanding sectoral and water-relevant general regulation, including, since the early 2000s, a growing body of EU Directives that together raise requirements (e.g. in drinking and surface water quality, and in financial performance and tariff caps) while narrowing the room for manoeuvring in decision-making. Stricter regulations on land use and spatial management, and nature protection, restrict options for sustained service delivery; many regulations prove simple add-ons rather than streamlined instruments reconciling opposing objectives and trade-offs. Illustratively, some utilities tend to prefer groundwater for supply augmentation over surface water because the former is less tightly regulated. Over the past two decades political and public scrutiny and interference have significantly increased by local governments seeking rents and local communities and vocal citizens expressing demands via informal (twitters) and formal (political) channels. Water Agencies in 2010 were close to being abolished as they were considered not cost-effective in the then-prevailing market liberalisation context. From the 1970s on water management has been recognised as knowledge-intensive: relying on novel technologies and a specialised work force. The sector has had to compete for qualified staff in a structurally tight labour market. Utilities and agencies now increasingly must align with an expanding set of municipal and regional policies such as on saving of water and

---

[4] For this survey, interviews were conducted of three CEOs of Utilities and three of Water Agencies. The respondents were selected based on their long-term exposure to the sector and their professional stature. The literature review concerned the main national trade journals, publications, and websites in 1986–2020.

Table 1 | Qualitative assessment of the drivers and pressures on water management from 1970 to 2030 as perceived by the leadership of representative drinking Water Utilities and Water Agencies (wastewater and flood management), The Netherlands.

Drivers and pressures	Reference situation 1970	Reference situation 1995	Reference situation 2020–2030
Laws and regulations (no. of)			
Utilities	2	8+3 EU Directives	11+4 EU Directives
Agencies	2	4+3 EU Directives	5+4 EU Directives
Political and public exposure			
Utilities	Low	Modest	High
Agencies	Low	Modest	Very high
Administrative challenge			
Utilities	Low	High (*introduction of water cycle*)	Very high (*local governments, circularity, decarbonisation*)
Agencies	Low		Very high (*region dev.*)
Staffing, know-how challenge			
Utilities	High	High	Very high
Agencies	Low	High (*esp. wastewater*)	Very high
Financing challenge			
Utilities	High: capital	Low	High
Agencies	Modest	High	High
Planning horizon			
Utilities	2–10 years	2–15 years	2–25 years
Agencies	2–10 years	2–15 years	2–25 years

other resources, decarbonisation, and circular economy. This imposes much closer collaboration and integration with other sectors and closer relationships with local and national-government administrations and industry. Whereas 50 years ago it was challenging to raise scarce capital for ambitious sectoral expansion investments, this constraint subsided in the 1990s (though less so for the agencies) but is now again a management pre-occupation as large physical assets such as sewers, pipelines and plants need to be replaced and measures against droughts put in place while tariffs are capped. Finally, the planning horizons have gradually extended to several decades in line with the growing capacity to forecast sectoral developments and the drive to manage assets and optimise investments. However, this generates significant uncertainty inherent to the forecasting models and the novel nature of many management measures.

Countries in development and in economic transition, on the other hand, arguably are facing a double challenge: complexity (and need for know-how) is growing and changes are rapid. The growth rate of complexity can be illustrated by observing that during the typical career of a water professional most countries are experiencing rapid growth in demography and prosperity. Simultaneously, their water resources and ecosystems are being exploited to, and now often beyond, their sustainability boundaries. A professional in such country, educated in the 1980s using textbooks published in the 1970s has a career spanning the period 1990–2030 and must manage the growing pressures on water in that period resulting from a demographic growth of 40–100%, the transition from a rural to an urban landscape, a 10-fold increase in tap connections, and households and industry turning 3–5 times as wealthy, each consuming 2–10 times more water. The fact that these countries often are heterogenous, comprising richer/urban and poorer/rural parts, adds to the complexity. In addition, they are facing the challenge of simultaneously addressing traditional water challenges (e.g. expanding service to all households) and the new ones of changing climate adaptation, rapid urban expansion, etc. They tend to rather rely on infrastructural approaches to manage supply but have also to start investing in the application of IWRM, institutional development and KCD. At the same time, the need for infrastructure remains large such as, in Africa, for water storage capacity.

Moving forward

The transitions in the physical and institutional realms possess three salient characteristics: complexity, uncertainty, and rapid change. Complexity refers to challenges that are associated with initial ignorance, deep uncertainty and potential contentiousness; cannot be mastered with straightforward measures (which distinguishes complex from complicated challenges); involve many different actors spanning various sectors and jurisdictions; are at least partly in continuing change so that, by the time part of the ignorance is uncovered, the circumstances have changed again; are transaction intensive; and necessitate an experiential discovery pathway of which the ultimate desired outcome is not guaranteed (Glouberman & Zimmerman, 2004; Snyder, 2013; Andrews *et al.*, 2017). This suggests that, looking forward, two main programme lines are unfolding. First, many challenges will project into an uncertain future and need an adaptive, iterative learning approach to uncover realistic pathways towards resolution. Second, the transition from static, bureaucratic approaches to 'organic' and dynamic ones able to manage the changes will increasingly necessitate institutional change and policy reform. The water sector needs to explore and equip itself with approaches and instruments to strengthen the knowledge base and institutional capacities required for addressing these transitions while complementing the traditional agendas that remain important.

THE NEXT CHALLENGE: KCD TO SUPPORT ADAPTIVE POLICY IMPLEMENTATION

Adaptive management formats

Whereas typical policies are normative, suggesting full knowledge of what needs to be achieved, several authors have proposed policies that are better able to deal with complexity, uncertainty and rapid change, to design and agree on new appropriate policies and implement them effectively. In uncertain situations, experts do not know, or the parties to a decision cannot agree upon (i) the external context of the system, (ii) how the system works and its boundaries, and/or (iii) the outcomes of interest from the system and/or their relative importance. Deep uncertainty also arises from actions taken over time in response to unpredictably evolving situations (Haasnoot *et al.*, 2013). Large-scale investment programmes, whether infrastructural (roads, energy, water) or social (education, health, etc.) typically span several decades and the design of the investments in the later part depends on fundamental early choices (path dependency) at a time when information is constrained most. The gap between available and needed knowledge will be smaller in countries with an extensive knowledge infrastructure and strong societal trust in knowledge and governance; adaptive management (AM) operates better in environments where cultural, financial and political space exists for 'learning' and mistakes. Developing countries, thus, may be at a disadvantage compared to richer economies.

Robust Decision-Making (RDM) (Lempert *et al.*, 2006; Bryant & Lempert, 2010) was proposed as a multiple-scenario evaluation framework for making decisions (in rich economies) on large infrastructural programmes with a large number of highly imperfect forecasts of the future. Rather than relying on improved point forecasts or probabilistic predictions, RDM describes many plausible futures relying heavily on stakeholder involvement, then helps analysts and decision-makers identify near-term actions that are robust across a wide range of futures – that is, actions that promise to do a reasonable job of achieving the decision-makers' goals compared to the alternative options, no matter what future comes to pass. Pahl-Wostl (2007) and Pahl-Wostl *et al.* (2007) propose AM to continually adjust water management decisions for Integrated Water Resources Management where positions of multiple stakeholders and longer-term consequences of decisions are uncertain. Kwakkel *et al.* (2016), Haasnoot *et al.* (2018) and Haasnoot & Warren (2019), also working mostly in rich economies on water investments to adapt to climate change, devised Dynamic Adaptive Policy Pathways (DAPPs) that continually adjust early adaptation management to environments that are complex, cannot be fully understood up-front, keep evolving and are sensitive to path-dependency. Andrews *et al.* (2017) propose for developing economies a Problem-Driven Iterative Adaptive Process (PDIA) to prepare effective longer-term development policies (irrespective of the sector) of which the outcomes are uncertain. These AM approaches are long-term iterative processes in which at the onset long-term goals are kept generic and vague but the goals of the imminent iteration well specified, based on the best current insights and making all assumptions explicit. By systematic M&E progress and failure, and assessing the assumptions after each step against expected outcome, uncertainty is reduced step-wise. Thus, the policy is gradually being implemented using best available knowledge and information is steadily accumulated to prepare future decisions. AM is a collection of approaches that help implement programmes or change a system by purposefully learning about the system. Surprisingly, only in the PDIA literature the need for KCD is mentioned explicitly; arguably, as the other AM approaches seem concerned with contexts in richer countries, they may assume knowledge and education systems already in place and sufficiently effective.

KCD support

AM approaches are knowledge-intensive and draw upon the Arena 6 KCD interventions, and thus also include those of the other Arenas. This intensity may put limits to the applicability of AM. For example, RDM rests on economic valuation of the alternative scenarios for which data may simply be lacking – typical for many developing countries. Because such policies always relate to longer-term programmes spanning decades, broad political support among decision-makers is paramount to ensure sustained commitment, also to the recognition that (partial) failures along the road are likely and represent valuable lessons shaping changes in the course of action that must be embraced. To enhance buy-in at the top, Lawrence & Haasnoot (2016) suggest to ensure opinion diversity by involving public sector actors from multiple governance levels and the private sector; structured interaction between scientists and stakeholders; and dedicated communication on uncertainty.

Once an AM policy is launched learning must be a constituent part, from the individual up to institutional levels (Fabricius & Cundill, 2014; Williams & Brown, 2018; Nikkels *et al.*, 2019; Pasanen & Barnett, 2019). AM requires structured M&E that is oriented towards both learning regarding effectiveness of the preceding steps and their assumptions, and accountability. Which M&E approaches are suitable for use in AM depends on the evidence and data needs to make informed adaptations, the type of programme for which it is used and available resources (Pasanen & Barnett, 2019). A useful format is the Results Framework applied by the World Bank, which identifies medium-term goals, indicators, short-term annual quantitative plan outputs, risks and recourse options. Similarly, the Theory of Change (ToC) is a mechanism to describe the set of assumptions that explain both the phases that lead to a long-term goal, and the connections between these activities and the programme's outcomes (Anderson, 2004). It describes the logic, principles and assumptions that connect what an intervention, service or programme does, and why and how it does it, with its intended results (Ghate, 2018). To be applied well, it demands an institutional willingness to be realistic and flexible in programming responses, both at the design stage and, more importantly, in implementation and performance management (Vogel, 2012).

Knowledge and capacity that are evidently needed here are the acquisition and management of data, the modelling and interpretation of data, and forecasting using mathematical simulation models and artificial intelligence. Rather than concerning a singular device or software, this refers to a system including education and training of modelers, scientists, technicians and decision-makers able to understand what data are required and how to manage it, and well-staffed institutions to finance, develop, maintain and update the specialised equipment.

As described above, once the policy or change programme is designed and agreed upon, its implementation raises challenges, especially in developing countries. The activities involved pertain to all KCD Arenas. Alaerts & Kaspersma (2009) categorise the competences that need to be available for AM and change programmes by the level where they need to be present (individual, organisational, and institutional (enabling) environment) and their nature: technical (in the sense of mono-disciplinary competence), managerial (to manage more complex systems), for governance, and for continuous learning. Based on Kaspersma (2013), Table 2 offers an indicative set of competences for AM and change programmes, dependent on context. Where competences are lacking or weak, KCD may draw from the instruments and tools described earlier.

Case study: adaptive change management in irrigation reform in Indonesia

Agriculture is by far the largest water user across the world; 250–300 million hectares of land are irrigated, supporting about 40% of global food production and several hundreds of millions rural households. Being capital intensive and partly a public function, initially it has often been managed by central technical administrations, raising concern because of moderate to low effectiveness in food productivity and water use, due to low farmer commitment, distorting subsidies and political interference. With rising incomes, many countries started reforms after the 1980s to enhance productivity as well as water security, and lower abstraction. Indonesia, a sprawling archipelago and lower middle-income country, depends heavily on irrigation and used the political and economic recalibration of 1999–2003 to launch a process to reform irrigation from strongly top-down management – with technical capacity, decision power and budgets in the central government – to a participatory-management model with substantial authority and budget shifted to local governments (about 250 districts/counties and 25 provinces) and Water User Associations (WUAs) of farmers. The policy was formulated in 1999–2003 involving all key stakeholders drawing from extensive field pilots and was implemented in the two decades thereafter through an adaptive, sequential approach with wide-scale technical and managerial (institutional) capacity development of local governments and WUAs. Lessons from the field drove stage-wise parallel policy and regulatory reform both at central and local-government levels. Indeed, farmers nor local governments had experience with or regulations for managing their own cropping and irrigation affairs, and at central policy level it was not clear which arrangements would eventually work best without risking embarking on an ineffectual course (Alaerts, 2020). Training of local government staff concerned notably participatory and agronomic practices, coaching of WUAs, water management, social marketing, articulation of local regulations, etc. Training of farmers concerned organisation of farmers, agricultural water use and allocation, agronomic practices, dike and canal maintenance, financial management, conflict resolution, etc. National government staff was concerned primarily with issuance of national regulations and M&E frameworks, and supervising very large irrigation systems. Relatively coherent longitudinal monitoring data sets are available on field-level practices; on economic returns on the six 'projects' of the World Bank and ADB that financed this reform and infrastructure; and on rice production at field, project and national levels. Institutional development was monitored by performance indicators. Well over three-quarters of the country are now applying the policy albeit with varying quality. Across the country teams of trainers/field coaches (recruited from NGOs and academia) were established; they had to follow regular

Table 2 | Framework categorising required indicative competences for AM and change programmes by level and nature.

	Individual level	Organisational level	Institutional environment level
Technical competence	Collect, process, and model the technical data needed that are the basis of AM approaches, either quantitative or more qualitative. Broader system knowledge.	Appropriate mix of technical knowledge and skills in place to implement AM. Modelling and forecasting expertise. Management of data and information.	Technical expertise and skills mix available in a broader setting, to many stakeholders. Systems for critical reflection and corroboration of knowledge.
Managerial competence	Project and programme management skills, leadership skills and attitudes, forward planning skills, personnel and organisational management skills. Able to appreciate and apply relevant traditional, indigenous knowledge.	Appropriate mix of technical and managerial knowledge and skills through conducive HR management. Able to operate with goals and objectives for AM. Able to deliver cost-effectively and timely. Able to appreciate and apply traditional, indigenous knowledge for sustainable solutions.	Minimal overlap, and complementarity between agencies, and mandate and size of agencies facilitate effective goal setting and task execution. Frameworks and incentives that enable engagement with and accomodation of different stakeholders and indigenous know-how.
Governance competence	Able to engage with stakeholders, and apply inclusiveness. Knowledge of sector and public administration institutions and how to bring AM forward. Able to behave ethically, e.g. in procurement, etc.	Able to project leadership and collaborate. Develop political consensus. Transparent and collaborative decision-making with multiple stakeholders, transparency in budgets and plans. Mechanisms for accountability and ethical behaviour.	Effective coordination among public sector actors, civil society and the private sector. Transparent and collaborative goal setting at societal level, with feed-back and accountability mechanisms to incentivise ethical behaviour.
Competence for continuous learning and innovation	Openness to learn, to acknowledge uncertainty and knowledge gaps. Peer-learning, networking, and ambition to improve water management	Facilitate learning by staff. Active coaching. M&E frameworks in place to support timely adaptation of strategies. Connect private and public sector and knowledge institutes. Peer-learning.	Open to review sector performance in line with AM. Able and willing to revise policies and planning when necessary. Foster inclusiveness in AM. Peer-learning.

(re-)training by Regional Experts, and fed back field experience. At field level water supply and allocation were reported more reliable and less wasteful, and conflicts better managed. Rice production increased significantly year after year after 2005, correlating with policy roll-out.

Figure 1 schematises the adaptive implementation of the change process which applied Arenas 1, 2, 3, 4, and 6 activities in a sequence of stages. In a conventional KCD process, first a problem is identified and knowledge and capacity needs assessed, after which the capacity is developed by a third party, and a result achieved; the institutional environment is assumed fixed. Here, on the other hand, the institutional environment (policies, and national and local regulations) were adjusted stage-wise and in parallel with the expanding body of field experience. Trainers/field coaches coached farmers and local governments and assisted in the structured monitoring and evaluation of outcomes. Academic institutes played an important role in education of new cohorts of professionals and in research. This process consumed much time: trainers had to be identified, trained and

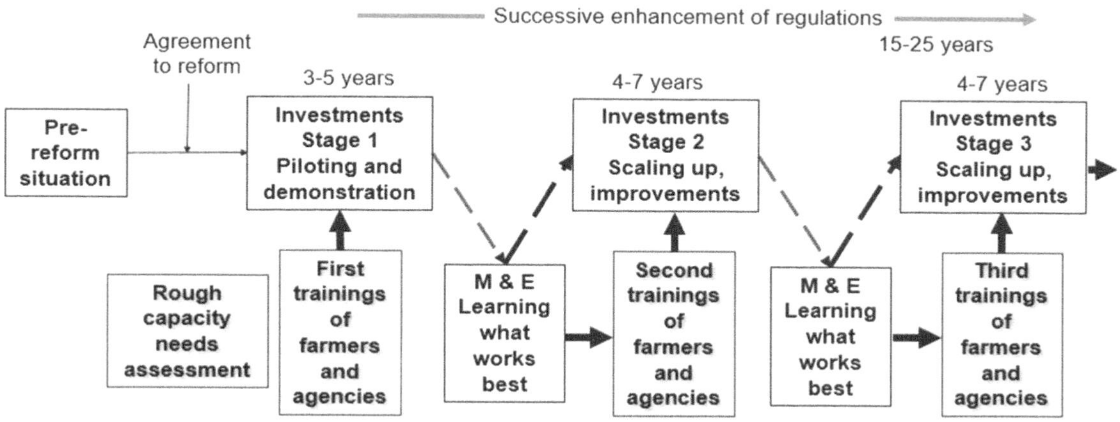

Fig. 1 | Schematisation of the stages of capacity development of farmers and local governments in an adaptive, sequential process; field lessons are fed back to inform the gradual enhancement and expansion of the national regulatory framework.

re-trained; work in remote areas took long preparation time; experiences needed to be shared and vetted; and with irrigation being seasonal, results could be assessed after one year only. Each stage took 3–7 years. As Indonesia is a large country with constrained resources, early policy roll-out was limited to a few selected districts/counties, with coverage scaled up after each stage to reach over three-quarters of the country after two decades. As institutional capacity develops slowly, full roll-out may require up to three decades. The analysis shows that in the context of 'wicked' policy implementation, not only up-front normative policy design matters, but equally importantly, the sequencing of the process of implementation allowing adaptation to physical, institutional and political realities. Still, to be successful a process must build on a credible realistic initial 'model'. This is especially true in developing and emerging economies with weaker institutional capacities and more fragmented governance structures.

The learning and KCD in the design and implementation stages conformed with theory: they helped inform negotiations and shape the coalitions at national and local levels (Sabatier & Jenkins-Smith, 1993) and facilitated the contextual interaction between the parallel processes (Saleth & Dinar, 2004; Bressers, 2007). They appeared critical for applying an adaptive approach (Pahl-Wostl, 2007; Andrews *et al.*, 2017). At the same time, these process guidelines must be case-specific as the theories do not provide operational guidelines. The policy roll-out confirmed that knowledge dissipation and ensuing administrative change are 'sticky' and get absorbed only slowly across the nation's institutions as described by Grindle (2004) and Nickson (2006). Senge (1990), Argyris (1992) and Nonaka & Takeuchi (1995) observed the same stickiness in corporations and relate it to mental and physical barriers to change processes. Structured and experiential learning help to address this (e.g. Alaerts, 1999; Easterby-Smith & Lyles, 2003; Andrews *et al.*, 2017).

Similar irrigation reform processes in other countries followed a less structured approach. In the Indian State of Andra Pradesh, the 1997 Irrigation Act shifted management authority from the centralised Irrigation Department to newly created WUAs, in a top-down reform (Oblitas & Peter, 1999). However, after 2000 the reform faltered; Reddy & Reddy (2005) attributed the failure to weak factual devolution, and absence of institutional capacity in WUAs and Irrigation Department with only cursory training on technical subjects for individuals, not for institutional change. The Indonesian process may have been more 'organic', adaptive and sequential supporting structured learning. Vietnam's irrigation sector has remained centralised through regional Irrigation Management

Companies under the Ministry of Agriculture and Rural Development. The large-scale Water Resources Assistance Project (2004–2012) applied participatory design but underperformed. Though the concept of participatory irrigation is by now accepted, the operational behaviour of the companies had not adapted. Importantly, the project ambitiously rolled out this new concept but failed to provide a phased design adaptation (IEG, 2019). In Morocco's irrigation reform (1900s–early 2000s) sequential reform was applied reasonably successfully (Doukkali, 2005). Morocco could build on earlier partial liberalisation in the 1970s with a next reform stage with the *Maroc Plan Vert* of 2008. Here, the iterative process worked well too, with continuous interaction between the implementation projects and formulation of the regulations and an 'adaptive' approach as opposed to an *ex ante* normative one as in Andra Pradesh. Its learning mode ensured that knowledge acquired from outside the country and from the field operations could be absorbed, and that the absence of know-how at the levels of local and central governments and among the farmers were remedied through field piloting, training and exchange between regulators and practitioners.

Saleth & Dinar (2000) recommend that a reform process take a pathway in which law, policy and organisation (administration) are addressed in a sequential yet consistent manner. A progressive step on one component needs to be followed by a consistent step in another, taking advantage of path dependencies, and providing mutual support, to identify and manage political resistance to the changes in a staged way while avoiding overreach. Uphoff (1986) and Ostrom (1990) also call for 'crafting of institutions' that responds to circumstances rather than imposing a normative approach. Studying irrigation in Australia, Nikkels *et al.* (2019) observe that irrigation planning is subject to unforeseeable changes in climate, technologies and societal preferences and, thus, requires DAPP-based management. They conclude that this is to be based on 'social learning', though do not elaborate on this.

THE ECONOMIC VALUE OF KCD

Knowledge and implementation capacity are critical input factors to economic and social development. Nevertheless, investment in sector-focused KCD is often modest and mostly confined to education and training (Arenas 1–3). Governments' readiness to invest depends on the expected economic rate of return (ERR). The economic value of a 'unit' of knowledge or capacity is hard to quantify as it is part of an economically productive activity and its impact hard to attribute causally to knowledge content. This value emerges through its application and depends on the type and context of the knowledge or capacity. However, for several situations valuation of knowledge/capacity is feasible by isolating the effect of a factor or through simulations. Nordhaus' seminal experiment (1996) on the value of innovation in lighting over the past three millennia suggests that over 99% of wealth creation derives from knowledge. The effect of most Arena 1 training to enhance specific skills can be causally linked to improved performance, and its economic value assessed. In other cases, especially where knowledge and capacity induce changes in the behaviour of many actors, and where other factors may affect the same outcome, attribution may be tenuous. In the case of education (Arena 3) the value of secondary and higher education has been estimated from the incremental salary enjoyed by graduates and the net benefit to society. In OECD countries, individuals with a secondary school degree tend to earn typically 23% more and those who attain tertiary education – most relevant for the knowledge-intensive water sector – an additional 50%. On average, the gross total return (net present value) of secondary and tertiary education is US$380,000 of which about 91,000 is public (or almost three times the amount of the public investment) and about 300,000 is private for men and 200,000 for women, representing a private ERR on the investment of about 12% for both sexes and about 10–11% for the public investment (OECD, 2011). For urban China, the economic, public return on schooling and higher education in the 1980s–1990s is estimated at 37–50% (Giles *et al.*, 2015). In Australia, the private return for a post-graduate degree is on average 47% and the public return 53% (Deloitte, 2016). These figures are averages across different education orientations and it is likely that disciplines such as IT, MBA, and finance would capture somewhat higher returns than water or civil engineering, which in turn may perform better than sciences such as ecology and microbiology. Thus, significant benefits are accruing to both individual and state. In developing economies, the value of education is arguably higher due to the relative scarcity of graduates but may in some instances be lower as shallow labour markets may not offer enough appropriate jobs to reward the education. In Arenas 2 and 4, knowledge and capacity in the corporate sector are embodied in knowledge products such as patents, goodwill, human capital and networks of clients and suppliers, which are valued for instance for the purpose of the valuation of firms and stocks. Many large corporations nowadays are deriving their value mostly from such intangible assets (Tobin, 1969; Cockburn & Griliches, 1987; Sveiby, 1997; Cummins & Bawden, 2010) though the net value of patents varies significantly across countries and sectors (EC, 2006). The return on private R&D and innovation (Arena 4) has been steadily growing between 1960 and the early 2000s to typically 50–100% as public return and 10–40% as private, but may be also reaching a ceiling in certain sectors; notably public R&D yields very variable outcomes (e.g. OECD, 1996). At least in Scandinavian countries a firm's value appears more related to 'organisational capital', or capacity, than to its R&D or patents (Rahko, 2014).

In the public water sector, especially in developing economies, valuing knowledge and capacity is more challenging. For Arena 2 change processes in water utilities, approximations have been proposed in qualitative terms by studying the incremental enhancement of administrative practices and KPIs (Pascual Sanz *et al.*, 2011; Mvulirwenande *et al.*, 2015). For an Arena 6 reform process, Alaerts (2020) derived the ERR from newly developed capacity supporting the implementation of Indonesia's national irrigation reform programme (see above). The projects financed physical rehabilitation plus institutional capacity development of both the local government's and national Irrigation Departments and of the WUAs. The return that can be imputed to capacity development was estimated at about 25%, or about two-thirds in the 36% of the aggregate return. In schemes without need for rehabilitation the ERR rose to 36%, generated solely by investment in capacity development. About one-quarter to one-third of the return remained imputable to the gains from improved institutional capacity. In addition, whereas the return on investment in infrastructure is confined to the lifetime of the physical asset, investment in institutions creates the foundation for further endogenous capacity development and better governance.

CONCLUSIONS

Knowledge and capacity have become more broadly accepted since the 1990s in water management and sector development, as being pivotal for defining goals and priorities and implementing the activities to achieve these. They have been demonstrated to cause impact and economic return. Yet, their application still tends to be often simplified to either general education or 'training'. Admittedly, comprehension of the full scope of KCD is challenging as it necessitates operational understanding of disciplines at the following three levels: (i) how the physical world is under threat and what action needs to be taken; (ii) how the institutional arrangements – policies, organisations, financial flows, etc. – need to be enhanced to achieve what is required under (i); and (iii) how educational, pedagogical and knowledge-management approaches can be applied to achieve the enhancement and change under (ii). Also in other sectors, knowledge and capacity are being placed at the core of development. Management sciences were first to discover the value of knowledge management, and many successful corporations started to roll out these approaches in the 1990s. In several countries the sectoral knowledge base and innovation became the subject of dedicated policies. Furthermore, international development theories have highlighted the deficiency of countries in their implementation capacity, to convert policy goals into effective action. Finally, in developed economies notably the health and environmental regulation communities have started developing an 'implementation science' to enhance their capacity to disseminate and embed new know-how faster.

The understanding and application of KCD have advanced substantially: (1) knowledge and capacity must be present in nested levels (individual, organisational, institutional, societal) which need to converge to cause effective action; (2) six arenas/contexts can usefully describe distinct categories of KCD initiatives; and (3) educational, pedagogical, and knowledge-management modes are outlined through which learning occurs and knowledge is imparted. However, too often KCD activities are disconnected from operational work, restricted to short-term project cycles or poorly implemented. The extensive literature on policy analysis tends to overlook the role of knowledge and capacity or treats it as a black box, perhaps because this literature rests mostly on experience in rich economies where ample knowledge and capacity are taken for granted. Many capacity-development guidelines tend to focus on the institutional change analysis, i.e. on 'what needs to be achieved' rather than on 'how to do it'. Yet KCD is a sticky, slow process meaning that it requires long-time engagement of many years or decades and adequate budget.

The water sector is facing growing, urgent challenges in the next three decades: the outstanding challenge of achieving the SDGs, and a new one of adapting to climate change and building a more resilient water-and-land system. To address these, three agendas can be discerned. First, current KCD activities and practices need to be scaled up and designed better based on a deeper understanding of effective approaches and instruments, notably applying longer-term programmatic frames and better embedding them in operational work. Second, more structured KCD must be adopted with medium- to long-term perspectives aiming to support change and reform processes at policy and organisational levels, to address the poor implementation of well-intentioned policies for want of institutional capacity. This challenge pertains to developing and rich countries alike. It calls for the discovery of what constitutes effective implementation through an iterative adaptive process and an implemenatation science. Third, as pressures on the water and land system across the globe are rising rapidly, effective policies must increasingly become pro-active, shaped by modelled forecasts; organisations and institutions will need to become enabled to change and adapt to future scenarios that are more complex, uncertain and evolving rapidly. Enhancing knowledge and (institutional) capacity for designing and implementing policies, and establishing 'learning organisations' – and, by extension, learning societies – are intensive and time-consuming processes, following iterative adaptive pathways, for which sustained political commitment and budget are preconditions.

Therefore we recommend the following. Foremost, as change and reform programmes require long-term and sustained political support, peer-learning among leaders (such as senior managers in the public and private sectors, and politicians) needs to be structured and facilitated, perhaps under the auspices of a new UN High-Level Panel of Government Leaders. Governments, water administrations and actors in civil society should prioritise analysis with a long-term horizon (e.g. for climate adaptation), develop capacity to set goals and achieve these, and apply learning approaches and instruments for implementation. However, while long-term programmatic strategies are advocated, the distinct supportive KCD activities should stay realistic and manageable in scope. Third, more dedicated research, knowledge sharing, and advocacy is required, involving practitioners and theoreticians from different disciplines and operating within a global impact monitoring framework; such a framework may develop from the existing framework for SDG6, but needs to be broader in scope and much more effective in shaping policy. To complete the investments and institutional changes in transitions takes 20–30 years; to be ready by 2050 they should be initiated now.

DATA AVAILABILITY STATEMENT

All relevant data are included in the paper or its Supplementary Information.

REFERENCES

ADB (2011). *Practical Guide to Capacity Development in A Sector Context.* Asian Development Bank, Manila.

Alaerts, G. (1999). Capacity building as knowledge management: purpose, definition and instruments. In *Water Sector Capacity Building: Concepts and Instruments* (Alaerts, G. J., Hartvelt, F. J. A. & Patorni, M.-F., eds). Balkema Publ., Rotterdam/Brookfield, VT.

Alaerts, G. J. (2009). Knowledge and capacity development (KCD) as tool for institutional strengthening and change. In *Water for a Changing World – Developing Local Knowledge and Capacity.* Alaerts, G. J. & Dickinson, N. L. (eds). CRC Press/Taylor&Francis, London.

Alaerts, G. (2020). Adaptive policy implementation: process and impact of Indonesia's national irrigation reform 1999–2018. *World Development 129*, 104880. https://doi.org/10.1016/j.worlddev.2020.104880.

Alaerts, G. (2022). Water, physically connected yet institutionally fragmented – investing in its strategies, asset classes and organizations. In *Financing Investment in Water Security.* Leflaive, X., Dominique, K. & Alaerts, G. (eds). Elsevier, Amsterdam/London.

Alaerts, G. & Kaspersma, J. (2009). Progress and challenges in knowledge and capacity development. In *Capacity Development for Improved Water Management.* Blokland, M. W., Alaerts, G. J., Kaspersma, J. M. & Hare, M. (eds). Taylor&Francis, London.

Alaerts, G., Blair, T. J. & Hartvelt, F. (1991). *A Strategy for Water Sector Capacity Building.* Int. Inst. Hydr. Envir. Engrg./UNDP, Delft/New York.

Anderson, A. (2004). *Theory of Change as a Tool for Strategic Planning: A Report on Early Experiences.* The Aspen Institute. Available at: https://www.aspeninstitute.org/publications/theory-change-tool-strategic-planning-report-early-experiences/ (accessed 5 February 2021).

Andrews, M., Pritchett, L. & Woolcock, M. (2017). *Building State Capability: Evidence, Analysis, Action.* Oxford University Press, Oxford.

Argyris, C. (1992). *On Organizational Learning.* Blackwell Publ., Malden, MA.

Bauer, M. S., Damschroder, L., Hagedorn, H., Smith, J. & Kilbourne, A. M. (2015). An introduction to implementation science for the non-specialist. *BMC Psychology 3*, 32. doi:10.1186/s40359-015-0089-9.

Bressers, H. (2007). *Contextual Interaction Theory and the Issue of Boundary Definition: Governance and the Motivation, Cognitions and Resources of Actors.* Institute for Governance Studies, University of Twente, The Netherlands.

Bressers, H. & de Boer, C. (2013). Conclusions. In *Water Governance, Policy and Knowledge Transfer.* de Boer, C., Vinke-Kruijff, J., Özerol, G. & Bressers, H. (eds). earthscan, Routledge, London.

Bryant, B. P. & Lempert, R. J. (2010). Thinking inside the box: a participatory, computer-assisted approach to scenario discovery. *Technological Forecasting & Social Change 77*, 34–49.

CBS (2012). *Monitoring of Top Sectors – First Measurement.* Central Bureau for Statistics, The Hague, The Netherlands.

Cockburn, I. & Griliches, Z. (1987). *Industry Effects and Appropriability Measures in the Stock Market's Valuation of R&D and Patents.* Report 2465, National Bureau of Economic Research, Cambridge, MA.

Cummins, J. & Bawden, D. (2010). Accounting for information: information and knowledge in the annual reports of FTSE 1–companies. *Journal of Information Science 36*(3), 283–305.

Deloitte (2016). *Estimating the Public and Private Benefits of Higher Education.* Australian Govt., Dept. of Education and Training, Canberra.

Doukkali, M. R. (2005). Water institutional reforms in Morocco. *Water Policy 7*, 71–88.

Easterby-Smith, M. & Lyles, M. A. (2003). *Handbook of Organizational Learning & Knowledge Management.* Wiley, Chichester, UK.

EC (2006). *Study on Evaluating the Knowledge Economy: What are Patents Actually Worth?* European Commission, DG Internal Market, Report ETD/2004/IM/E3/77, Brussels.

EuropeAid (2005). *Institutional Assessment and Capacity Development.* EuropeAid, European Commission, Brussels.

Fabricius, C. & Cundill, G. (2014). Learning in adaptive management: insights from published practice. *Ecology and Society 19*(1), 8–16. Available at: https://www.jstor.org/stable/26269492 (accessed 10 December 2020).

Famiglietti, J. S. (2014). The global groundwater crisis. *Nature Climate Change 4*, 945–948. Available at: www.nature.com/natureclimatechange (accessed March 2021).

Ghate, D. (2018). Developing theories of change for social programmes: co-producing evidence-supported quality improvement. *Palgrave Communications 4*(1), 1–13. doi:10.1057/s41599-018-0139-z.

Gidron, B. & Bar-Almog, M. (2010). *Policy Initiatives Towards the Third Sector in International Perspective.* Springer, Berlin.

Giles, J., Park, A. & Wang, M. -Y. (2015). *The Great Proletarian Cultural Revolution, Disruptions to Education, and the Returns to Schooling in Urban China.* Hong Kong University of Science and Technology IEMS, Hong Kong, Working Paper No. 2015-21.

GiZ (2015). *Cooperation Management for Practitioners. Managing Social Change with Capacity WORKS.* Springer-Gabler, Wiesbaden.

Gleeson, T., Cuthbert, M., Ferguson, G. & Perrone, D. (2020). Global groundwater sustainability, resources, and systems in the anthropocene. *Annual Review of Earth and Planetary Sciences 48*, 431–463. https://doi.org/10.1146/annurev-earth-071719-055251 (accessed January 2021).

Glouberman, S. & Zimmerman, B. (2004). Complicated and complex systems: what would successful reform of medicare look like? In *Romanow Papers: Changing Healthcare in Canada*, Vol. 2. Forest, P. G., Marchildon, G. P. & McIntosh, T. (eds). University of Toronto Press, Toronto, p. 31.

Grindle, M. S. (2004). *Despite the Odds: Contentious Politics and Educational Reform.* Princeton University Press, Princeton, NJ.

Guilmette, J. -H. (2009). *Power of Peer Learning: Networks and Development Cooperation.* IDRC, Ottawa, Ontario. hdl:10625/28624.

Haasnoot, M. & Warren, K. (2019). Dynamic adaptive policy pathways (DAPP). In *Decision Making Under Deep Uncertainty.* Marchau, V. A. W. J., Walker, W. E., Bloemen, P. J. T. M. & Popper, S. W. (eds). Springer Nature, Cham, Switzerland. doi:10.1007/978-3-030-05252-2_4.

Haasnoot, M., Kwakkel, J. H., Walker, W. E. & ter Maat, J. (2013). Dynamic adaptive policy pathways: a method for crafting robust decisions for a deeply uncertain world. *Global Environmental Change 39*, 485–498.

Haasnoot, H., van Aalst, M., Rozenberg, J., Dominique, K., Matthews, J., Bouwer, L. M., Kind, J. & Poff, N. L. (2018). Investments under non-stationarity: economic evaluation of adaptation pathways. *Climatic Change.* https://doi.org/10.1007/s10584-019-02409-6.

Hering, J. (2018). Implementation science for the environment. *Environmental Science & Technology 52*, 5555–5560. doi:10.1021/acs.est.8b00874.

Herrera-García, G., Ezquerro, P., Tomás, R., Béjar-Pizarro, M., López-Vinielles, J., Rossi, M., Mateos, R. M., Carreón-Freyre, D., Lambert, J., Teatini, P., Cabral-Cano, E., Erkens, G., Galloway, D., Hung, W. -C., Kakar, N., Sneed, M., Tosi, L., Wang, H. M. & Ye, S. J. (2021). Mapping the global threat of land subsidence. *Science 6524*, 34–36.

IEG (2019). *Vietnam. Water Resources Assistance Project.* Independent Evaluation Group, World Bank, Washington, DC.

IPCC (2021). Climate change 2021: the physical science basis. In *Contribution of WG I to the Sixth Assessment Report of the Intergovernmental Panel on Climate Change* (Masson-Delmotte, V. *et al.*, eds). Cambridge University Press, Cambridge/New York.

IPCC (2022). Climate change 2022: impacts, adaptation and vulnerability. In *Contribution of WG II to the Sixth Assessment Report of the Intergovernmental Panel on Climate Change* (Partner, H.-O. *et al.*, eds). World Meteorological Organization, Geneva, and United Nations Environmental Programme, Nairobi.

Jensen, O. (2019). Designing effective water policy: capacity and effectiveness of reforms in developing countries. *Policy and Society 38*(1), 77–95. doi:10.1080/14494035.2018.1513438.

Kaspersma, J. (2013). *Competences in Context: Knowledge and Capacity Development in Public Water Management in Indonesia and The Netherlands.* Routledge/CRC Press, Milton Park, UK.

Keijzer, N., Spierings, E., Phlix, G. & Fowler, A. (2011). *Bringing the Invisible Into Perspective.* European Centre for Development Policy Management, Maastricht.

Kingdon, J. W. (1995). *Agendas, Alternatives, and Public Policies*, 2nd edn. Longman, New York.

Kolb, D. A. (1984). *Experiential Learning: Experience as the Source of Learning and Development.* Prentice Hall, Englewood Cliffs, NJ.

Kuhn, T. S. (1962). *The Structure of Scientific Revolutions.* University of Chicago Press, Chicago.

Kwakkel, J. H., Haasnoot, M. & Walker, W. E. (2016). Comparing robust decision-Making and dynamic adaptive policy pathways for model-based decision support under deep uncertainty. *Environmental Modelling and Software 86*, 168–183. doi:10.1016/j.envsoft.2016.09.017.

Lawrence, J. & Haasnoot, M. (2016). What it took to catalyse uptake of dynamic adaptive pathways planning to address climate change uncertainty. *Environmental Science and Policy 68*, 47–57. doi:10.1016/j.envsci.2016.12.003.
Lempert, R. J., Groves, D. G., Popper, S. W. & Bankes, S. C. (2006). A general, analytic method for generating robust strategies and narrative scenarios. *Management Science 52*(4), 514–528.
Lombardo, M. M. & Eichinger, R. W. (1996). *The Career Architect Development Planner*. Lominger, Minneapolis.
Mangin, M. M. & Dunsmore, K. (2014). How the framing of instructional coaching as a lever for systemic or individual reform influences the enactment of coaching. *Educational Administration Quarterly 51*(2), 179–213.
McGrath, R. G. (2011). The World is More Complex than it used to be. Harvard Business Review, June–August, 2011. Available at: https://hbr.org/2011/08/the-world-really-is-more-compl.html (accessed December 2020).
Meijerink, S. & Huitema, D. (2009). Understanding and managing policy transitions: a policy science perspective. In *Water Policy Enterpreneurs: A Research Companion to Water Managers Around the Globe*. Meijerink, S. & Huitema, D. (eds). Edward Elgar Publ., Cheltenham.
Metze, M. (2008). *Veranderend getij (Shifting Tides)*. Balans, Amsterdam.
Ministry of Economic Affairs, Agriculture and Innovation (2011). *Water Deserves It. Advice on the Top Sector Water*. Netherlands Govt., The Hague (in Dutch).
Morgan, P. (1993). *Capacity Building: An Overview. Workshop on Capacity Development*. Institute on Governance, CIDA, Ottawa.
Mugisha, S. & Brown, A. (2010). Patience and action pays: a comparative analysis of WSS reforms in three East African cities. *Water Policy 12*, 654–674.
Mugisha, S., Berg, S. V. & Muhairwe, W. T. (2007). Using internal incentive contracts to improve water utility performance: the case of Uganda's NWSC. *Water Policy 9*(3), 271–284. doi:10.2166/wp.2007.010.
Mvulirwenande, S., Wehn, U. & Alaerts, G. (2015). Closing the learning cycle in capacity development impact evaluation: a two-step approach. In *Proc. 38th WEDC International Conference*, Loughborough University, Loughborough.
Nickson, A. (2006). Public sector management reform in Latin America. In *Public Sector Reform in Developing Countries*. Bangura, Y. & Larbi, G. A. (eds). Palgrave Macmillan, New York, NY.
Niedenthal, P., Krauth-Gruber, S. & Ric, F. (2006). *Psychology of Emotion: Interpersonal, Experiential and Cognitive Approaches*. Psychology Press, New York/Hove.
Nikkels, M. J., Kumar, S. & Meinke, H. (2019). Adaptive irrigation infrastructure – linking insights from human-water interactions and adaptive pathways. *Current Opinion in Environmental Sustainability 40*, 37–42. doi:10.1016/j.cosust.2019.09.001.
Nonaka, I. & Takeuchi, H. (1995). *The Knowledge-Creating Company*. Oxford University Press, New York/Oxford.
Nordhaus, W. D. (1996). Do real-output and real-wage measures capture reality? The history of lighting suggests not. In *The Economics of New Goods*. Bresnahan, T. & Gordon, R. (eds). University of Chicago Press, Chicago, IL.
North, D. C. (1990). *Institutions, Institutional Change and Economic Performance*. Cambridge University Press, Cambridge, UK.
NRC (2020). Zo lang de minister maar niet struikelt (Provided the Minister doesn't trip). NRC Newspaper, *Rotterdam*, 21 and 22-23 February 2020 (in Dutch).
Oblitas, K. & Peter, J. R. (1999). *Commencing Irrigation Sector Reforms Through Management Transfer to Farmers: The Case of Andra Pradesh, India*. Technical Paper. The World Bank, Washington, DC.
OECD (1996). *The Knowledge-Based Economy*. Organisation for Economic Co-operation and Development, Report OCDE/GD(96)102, Paris.
OECD (2011). *Education at a Glance 2011*. Organisation for Economic Co-operation and Development, Paris.
OECD (2012). *OECD Environmental Outlook to 2050*. Organisation for Economic Co-operation and Development, Paris.
OECD (2014). *Water Governance in The Netherlands. Fit for the Future?* Organisation for Economic Co-operation and Development, Paris.
OECD (2017). *The Land-Water-Energy Nexus: Biophysical and Economic Consequences*. Organisation for Economic Co-operation and Development, Paris.
OECD (2019a). *Governance as an SDG Accelerator*. Organisation for Economic Co-operation and Development, Paris.
OECD (2019b). *Biodiversity: Finance and the Economic and Business Case for Action*. Organisation for Economic Co-operation and Development, Paris.
OECD (2021). *Forum on Green Finance and Investment 2021*. Organisation for Economic Co-operation and Development, Paris. Available at: https://oecd-events.org/green-finance/sessions?searchtext=&page=1 (accessed 1 November 2021).

O'Rourke, E. (1992). The international drinking water supply and sanitation decade: dogmatic means to a debatable end. *Water Science and Technology 26*(7–8), 1929–1939. https://doi.org/10.2166/wst.1992.0638.

Ostrom, E. (1990). *Governing the Commons: The Evolution of Institutions for Collective Action*. Cambridge University, New York.

Pahl-Wostl, C. (2007). Transitions towards adaptive management of water facing climate and global change. *Water Resources Management 21*, 49–62.

Pahl-Wostl, C., Craps, M., Dewulf, A., Mostert, E., Tabara, D. & Taillieu, T. (2007). Social learning and water resources management. *Ecology and Society 12*(2), 5.

Pasanen, T. & Barnett, I. (2019). *Supporting Adaptive Management – Monitoring and Evaluation Tools and Approaches*. Working paper 569, Institute of Development Studies, London. Available at: https://cdn.odi.org/media/documents/odi-ml-adaptivemanagement-wp569-jan20.pdf.

Pascual Sanz, M., Schouten, M., Alaerts, G. & van Tulder, R. (2011). Appraising water operator's capacity to reduce non-revenue water. The case of Blantyre Water Board. In *6th Spec Conf on Efficient Use and Management of Water. Water Demand Management: Challenges and Opportunities*, 29 March–2 April 2011. International Water Association, Dead Sea, Jordan.

Polanyi, M. (1958). *Personal Knowledge: Towards a Post-Critical Philosophy*. Routledge & Kegan Paul, London.

Portmann, F. T. (2013). Impact of climate change on renewable groundwater resources: assessing the benefits of avoided greenhouse gas emissions using selected CMIP5 climate projections. *Environmental Research Letters 8*(2), 024023(14pp). doi:10.1088/1748-9326/8/2/024023.

Pritchett, L., Samji, S. & Hammer, J. (2013). *It's All About MeE: Using Structured Experiential Learning ('e') to Crawl the Design Space*. Center for Global Development Working Paper 322, Washington, DC. Available at: http://www.cgdev.org/publication/its-all-about-mee (accessed July 2020).

Rahko, J. (2014). Market value of R&D, patents, and organizational capital: Finnish evidence. *Economics of Innovation and New Technology 23*(4), 353–377. doi:10.1080/10438599.2013.864923.

Reddy, V. R. & Reddy, P. P. (2005). *How Participatory is Participatory Irrigation Management (PIM)? A Study of Water User Associations (WUAs) in Andra Pradesh*. Centre for Social and Economic Studies, Hyderabad.

Rotmans, J., Kemp, R. & van Asselt, M. (2001). More evolution than revolution: transition management in public policy. *Foresight 3*(1), 15–31.

Sabatier, P. A. & Jenkins-Smith, H. C. (1993). *Policy Change and Learning: An Advocacy Coalition Approach*. Westview Press, Boulder, CO.

Saleth, R. M. & Dinar, A. (2000). Institutional changes in global water sector: Trends, patterns and implications. *Water Policy* 2, 175–199.

Sadoff, C. W., Hall, J. W., Grey, D., Aerts, J. C., Ait-Kadi, M., Brown, C., Cox, A., Dadson, S., Garrick, D. & Kelman, J. (2015). *Securing Water, Sustaining Growth: Report of the GWP/OECD Task Force on Water Security and Sustainable Growth*. University of Oxford, Oxford.

Saleth, R. M. & Dinar, A. (2004). *The Institutional Economics of Water: A Cross-Country Analysis of Institutions and Performance*. Edward Elgar, Cheltenham.

Schein, E. H. (1978). *Career Dynamics: Matching Individual and Organizational Needs*. Addison-Wesley, Reading, MA.

Schewe, J., Heinke, J., Gerten, D. & Kabat, P. (2014). Multi-model assessment of water scarcity under climate change. *Proceedings of the National Academy of Sciences of the United States of America 111*(9), 3245–3250. doi:10.1073/pnas.1222460110.

Senge, P. (1990). *The Fifth Discipline: The Art and Practice of the Learning Organization*. Currency Doubleday, New York, NY.

Snyder, S. (2013). *The Simple, the Complicated, and the Complex: Educational Reform Through the Lens of Complexity Theory*. Organisation for Economic Co-operation and Development, Education Working Paper 96, Paris.

Sveiby, K. E. (1997). *The New Organizational Wealth: Managing and Measuring Knowledge Based Assets*. Berrett Koehler, San Francisco, CA.

Sveiby, K. E. (2001). A knowledge-based theory of the firm to guide in strategy formulation. *Journal of Intellectual Capital 2*(4), 344–358. https://doi.org/10.1108/14691930110409651.

Tobin, J. (1969). A general equilibrium approach to monetary theory. *Journal of Money, Credit and Banking 1*(1), 15–29.

Turner, S. W. D., Hejazi, M., Yonkofski, C., Kim, S. H. & Kyle, P. (2019). Influence of groundwater extraction costs and resource depletion limits on simulated global nonrenewable water withdrawals over the twenty-first century. *Earth's Future 7*, 123–135. https://doi.org/10.1029/2018EF001105.

Uhlenbrook, S. & de Jong, E. (2012). T-shaped competency profile for water professionals of the future. *Hydrology and Earth System Sciences 16*, 3475–3483. https://doi.org/10.5194/hess-16-3475-2012.

UNDP (2007). *Supporting Capacity Development – The UNDP Approach*. United Nations Development Programme, Bureau for Development Policy, New York, NY.

Uphoff, N. (1986). *Improving International Irrigation Management with Farmer Participation: Getting the Process Right*. Westview Press, Boulder, CO.

Van der Brugge, R., Rotmans, J. & Loorbach, D. (2005). The transition in Dutch water management. *Regional Environmental Change 5*, 164–176.

Van der Ent, R. J., Savenije, H., Schaefli, B. & Steele-Dunne, S. C. (2010). Origin and fate of atmospheric moisture over continents. *Water Resources Research 46*, W09525. doi:10.1029/2010WR009127.

Van der Esch, S., ten Brink, B., Stehfest, E., Bakkenes, M., Sewell, A., Bouwman, A., Meijer, J., Westhoek, H. & van den Berg, M. (2017). *Exploring Future Changes in Land use and Land Condition and the Impacts on Food, Water, Climate Change and Biodiversity: Scenarios for the Global Land Outlook*. PBL Netherlands Environmental Assessment Agency, The Hague.

Vogel, I. (2012). *Review of the Use of 'Theory of Change' in International Development*. UK Department for International Development, London. Available at: https://www.researchgate.net/publication/259999430_Review_of_the_Use_of_'Theory_of_Change'_in_International_Development (accessed 5 February 2021).

Williams, B. K. & Brown, E. D. (2018). Double-loop learning in adaptive management: the need, the challenge, and the opportunity. *Environmental Management 62*(6), 995–1006. doi:10.1007/S00267-018-1107-5.

World Bank (1998). *Knowledge for Development. World Development Report 1998/99*. The World Bank, Washington, DC.

World Bank (2016). *High and Dry: Climate Change, Water and the Economy*. The World Bank, Washington, DC.

World Bank (2018). *Managing Knowledge, Learning and Data for Operations*. The World Bank, DEC Knowledge Management Group, Washington, DC. Available at: https://pubdocs.worldbank.org/en/248681543420897102/DECKM-brochure.pdf (accessed January 2021).

WWAP (2018). *World Water Development Report 2018*. United Nations World Water Assessment Programme, UNESCO, Paris.

WWF (2020). In *Living Planet Report 2020 - Bending the Curve of Biodiversity Loss*. Almond, R. E. A., Grooten, M. & Petersen, T. (eds). World Wildlife Fund, Gland, Switzerland.

First received 17 December 2021; accepted in revised form 19 April 2022. Available online 13 May 2022

© 2022 The Authors From Knowledge and Capacity Development to an Implementation Science: doi: 10.2166/wp.2021.264

Social change innovations, citizen science, miniSASS and the SDGs

Jim Taylor [a,*], Mark Graham[b], Adrienne Louw[b], Ayanda Lepheana[b], Bonani Madikizela[c], Chris Dickens[d], Deborah V. Chapman[e] and Stuart Warner [e]

[a] University of KwaZulu-Natal, Centre for Water Resources Research, Pietermaritzburg, South Africa
[b] GroundTruth, Hilton, South Africa
[c] Water Research Commission, Pretoria, South Africa
[d] International Water Management Institute, Colombo, Sri Lanka
[e] UN Environment Programme GEMS/Water Capacity Development Centre, School of Biological, Earth & Environmental Sciences and Environmental Research Institute, University College Cork, Cork, Ireland
*Corresponding author. E-mail: jimtaylor835@gmail.com

JT, 0000-0002-5461-1582; SW, 0000-0002-2626-2784

ABSTRACT

The United Nations Sustainable Development Goals (SDGs) describe a course of action to address poverty, protect the planet and ensure prosperity for all (https://sdgs.un.org/goals). More specifically, SDG 6 clarifies how water quality, quantity and access are crucial to human well-being, and yet human activities are compromising water resources through over-exploitation, pollution, as well as contributing to the spread of disease. Globally aquatic ecosystems are highly threatened and concerted efforts by governments and civil society to 'turn the situation around' are simply not working. Human-created problems require human-centred solutions and these require different ways of thinking and acting to those behaviour patterns that are contributing to the challenges. In this paper, we first consider *causal* approaches to attitude change and behaviour modification that are simply not working as intended. We then explore *enabling* responses such as citizen science and co-engaged action learning as more tenable alternatives. SDG 6 has a focus on clean water and sanitation for all. The SDGs further clarify how the extent to which this goal can be realized depends, to a large extent, on stakeholder engagements and education. Through stakeholder engagements and educational processes, people can contribute towards SDG 6 and the specific indicator and target in SDG 6.b – Stakeholder participation. Following a three-year research process, that investigated a wide range of participatory tools, this paper explores how the Stream Assessment Scoring System (miniSASS; www.minisass.org) can enable members of the public to engage in water quality monitoring at a local level. The paper continues to demonstrate how miniSASS can contribute to the monitoring of progress towards Sustainable Development Goal Target 6.3, by providing a mechanism for data collection indicator 6.3.2. miniSASS is proving popular in southern Africa as a methodology for engaging stakeholder participation in water quality monitoring and management. The technique costs very little to implement and can be applied by children and scientists alike. As a biomonitoring approach, it is based on families of macroinvertebrates that are present in most perennial rivers of the world. The paper concludes by describing how useful the miniSASS technique can be for addressing data gaps for SDG 6.3.2 reporting, and that it can be applied in most regions of the world.

Key words: Biomonitoring, Citizen science, miniSASS, SDG6, Water quality

HIGHLIGHTS

- The Sustainable Development Goals (SDGs) are a globally approved policy and action process.
- This paper demonstrates how citizen-derived data through the stream assessment scoring system (miniSASS) can be used for SDG 6 reporting on water quality.

- It explores the expansion of miniSASS to collect water quality data globally.
- miniSASS supports policy by mobilizing society to engage with water issues.

INTRODUCTION

Conventional wisdom continues to fail us

Conventional wisdom approaches, which often emphasize attitude change with the assumption that behavioural practices will follow have not proved tenable (Beck, 1992, 1995, 1997; Kemmis & Mutton, 2012). Fien (2003) emphasizes that 'among the most successful [environmental education] programmes are those that avoid the belief that awareness leads to understanding, understanding leads to concern, and concern motivates the development of *skills and action* (our italics)' (Fien, 2003). Causal, linear, top-down or centre-to-periphery approaches that assume behaviour change, following awareness raising, often facilitate a power-gradient from those who feel they know to those who they feel ought to know. This rational logic continues to assume that once informed, the others, often described as a target group[1], will change accordingly! As reported above, this rational change process fails to meet expectations and, at times, may even alienate the very people it is seeking to change (Taylor, 2010).

The sustainable development goals and SDG 6

SDG 6 has a focus on clean water and sanitation for all. The SDGs further clarify how the extent to which this goal can be realized depends, to a large extent, on stakeholder engagements and education. Through stakeholder engagements and educational processes, people can contribute towards SDG 6 and the specific indicator and target in SDG 6.b – Stakeholder participation, namely:

> 'Support and strengthen the participation of local communities in improving water and sanitation management'

The United Nations (UN) are explicit as to why stakeholder participation is important, but a key research question is how this participation can be meaningful and engaging, particularly the aspect of improving water management:

> *'Target 6.b aims for the participation of local communities in water and sanitation planning and management, which is essential for ensuring that the needs of all people are being met. The involvement of relevant stakeholders is further necessary to ensure: that the technical and administrative solutions decided upon are suitable for specific socioeconomic contexts, the full understanding of the impacts of a certain development decision and the encouragement of local ownership of the solutions when implemented (to ensure sustainability over time). Target 6.b supports the implementation of all SDG 6 targets (targets 6.1–6.6 and 6.a) by promoting the meaningful involvement of local communities, which is also a central component of IWRM'*[2].

A detailed three-year research process, supported by the Water Research Commission, investigated a wide range of public participation approaches to water quality issues and clarified how citizen science can play a

[1] The use of the term target group, a metaphor from military operations, emphasizes the causal tradition of 'getting the message across.' It is therefore difficult to apply in a more inclusive, dialogue-centred manner. By classifying people as 'the other' it militates against an opportunity for relationship building, mutual learning, deliberation and a commitment to building understanding and human dignity.

[2] https://www.sdg6monitoring.org/indicators/target-6b/

meaningful role in water-related issues (Graham & Taylor, 2018). Of all the citizen science tools that were reviewed, miniSASS stood out as a relatively easy technique to use, that can be applied at virtually no cost, the results are immediately available and no laboratory is necessary to develop or interpret the findings.

Stakeholder participation through citizen science

Citizen science and co-engaged action learning (Bonney *et al.*, 2009; Pocock, *et al.*, 2014; O'Donoghue *et al.*, 2018) are, however, showing the way to more inclusive, enabling and effective social change processes. This work deepens the understanding of water issues in a practical and applied manner and enables actions for more sustainable practices (Graham & Taylor, 2018). Here the democratization of science engages people who often become proud and eager participants in building understanding and working for more sustainable practices rather than being the passive recipients of knowledge from others.

For challenges as complex as water management and ecological infrastructure[3], the importance of context and the involvement of those participating is crucial. Many of these issues and problems require the integration of knowledge from the natural and social sciences, as well as economics, and there is rarely a single 'silver bullet' solution. As Bhaskar *et al.* (2010) states, 'exemplifying the triangular relationship of critical realism, interdisciplinarity and complex (open-systemic) phenomena' is needed for the investigation of such wicked problems as water resources management.

As stated above 'The democratisation of science, through citizen science processes, supported by practical and accessible 'tools of science,' is one area of work that is showing encouraging results' (Graham & Taylor, 2018, page V). Building collective capacity in the context of resource-based risks and uncertainty points towards the broad field of social learning (Wals, 2007) as a useful component of effective social change. These ideas and concepts form the basis of this paper which advocates an 'enabling' orientation – rather than a traditional 'causal' approach (i.e. seeking to cause a change in others through top-down methodologies) (Taylor, 2014).

miniSASS, biomonitoring and Google Earth

We now explore miniSASS as a key enabling tool that may have global relevance in both mobilizing people, popularizing the Sustainable Development Goals (SDGs) and helping provide accessible biomonitoring data, at virtually no cost to the user, towards SDG indicator 6.3.2. This perspective on citizen science is mirrored in a seminal paper by Fritz *et al.* (2019) on citizen science and the United Nations SDGs. In their paper, Fritz *et al.* develop a road-map on how citizen science can support the SDGs. In keeping with this theme, we are exploring miniSASS as a complementary, citizen science orientated, research tool.

What is the stream assessment scoring system (miniSASS)?

miniSASS is a simple tool which can be used by anyone to monitor the health of a river. One simply collects a sample of macroinvertebrates (small organisms large enough to be seen with the naked eye) from a natural river or stream, and depending on which groups are present, one can calculate a River Health Index for the river. This score helps classify the health, or ecological condition of the river, ranging across five categories from natural (blue) to very poor (purple). The results can then be recorded on the miniSASS website Google Earth layer (www.minisass.org). This database is effectively a 'Living data' system where further data can continuously be added. Through miniSASS, one can learn about rivers, monitor their water quality, explore the drivers of water quality deterioration, and, of course, take action to improve the quality of the streams and rivers.

---

[3] Understanding the importance of ecological infrastructure (EI) which may be defined as *functioning ecosystems that produce and deliver valuable services to people* (Jewitt *et al.*, 2020), may be a broader and more useful term to help understand catchment management processes.

Challenges to be considered when applying miniSASS

As with any scientific enquiry processes, a number of challenges must be addressed when applying miniSASS, these include:

- Participants must locate a local stream or river and be prepared to go into the water to locate the organisms.
- The organisms are not always easy to catch for identification purposes. A small net, such as one used for a gold-fish pond, can be helpful here.
- Once collected, the organisms need to be identified. This is not always easy. The miniSASS website, at www.minisass.org, does, however, provide a simple dichotomous key through which the users can key-out the organism they are seeking to identify.
- In any citizen science, endeavour issues of scientific accuracy are a concern. By offering people training courses in miniSASS, this issue is alleviated to some extent. Training can be undertaken online using a simple, instructional video from the miniSASS website referred to above. The validity of the data will be taken up further in the Results and Discussion section.

Where did miniSASS come from?

miniSASS (Graham *et al.*, 2004) was derived from the more rigorous South African Scoring System (SASS), a biomonitoring method for evaluating river health (Dickens & Graham, 2002; Dallas & Day, 2004). The SASS method requires the ability to identify up to 90 different aquatic invertebrate families and thus a high level of training is required. There was, therefore, a need for a simpler, more user-friendly tool for biomonitoring that would still yield reliable water quality/river health data. This need gave rise to the development of miniSASS. The development process involved reducing the taxonomic complexity of SASS by creating broad groupings of invertebrates that could act as surrogates for the complete suite of SASS taxa. A rigorous statistical evaluation of a large volume of data from the SASS method was conducted in order to determine whether the miniSASS method would yield viable results, similar to those derived from SASS. This evaluation assessed data collected over a wide geographic and water quality range and provided indications that miniSASS is suitable for predicting SASS scores and is thus an appropriate indicator of biological water quality (Graham *et al.*, 2004).

The strength of miniSASS has been evident for several years. Its use is widely applied in South Africa, where it was initially developed, although it has been successfully used in many other southern African countries. It has been effectively applied in India (in the Himalaya mountains at over 18,000 feet in altitude), in Vietnam, Canada (where the ambient air temperature was –20°C), Germany and Brazil. GroundTruth, an environmental engineering company, verifies the incoming data and has worked with the Water Research Commission to support this development through the www.minisass.org website. A topical feature of this website is the number of recent additions of data, as shown on the right-hand side of the screen. miniSASS is also found on the Water Research Commission's website www.wrc.org.za.

RESULTS AND DISCUSSION

The existing and future potential of miniSASS as a global citizen science monitoring tool suitable for SDG 6.3.2

Substantial development of miniSASS has taken place since its early inception. This work includes the development of an online portal, or website described above, that allows participants, ranging from school children to NGOs and other organizations, to upload their self-collected data. Although there may be concern about the validity of these data, largely due to a lack of training of those collecting it, this concern is offset by the fact that the data are 'crowd-sourced', allowing for the shear bulk of evidence to tell a valid story, perhaps even more reliably than the more sparsely collected 'official' data. Holt *et al.* (2013) describe and demonstrate the potential of citizen science data. They show that, despite biases, it is possible to statistically clean citizen science data so that it matches the quality of professional sampled data. Citizen-derived data can also cover huge geographical regions, whereas professionals can only cover small areas (due to the cost).

How universal is miniSASS?

Because many macroinvertebrates have part of their life cycle in the water and part in the air, they have been able to move vast distances, carried by water, wind and by ducks and other vectors, and are thus found spread across the world[4]. However, others, such as some crustaceans, have not moved so far and tend to be more confined. For this reason, macroinvertebrate samples collected across the world have many similarities, especially at the taxonomic level of order or family, while at a local level, there is plenty of division into local genera and species. For this reason, indices that are based on the higher taxonomic level of order and family are more globally applicable. This was validated, in South Africa, and was a large part of the success of the SASS index (Dickens & Graham, 2002) which is based largely on family identification, making the index inexpensive and thus affordable for routine and large-scale monitoring.

An index that is to be applied at a global level could be improved with local adaptations. This can be done, not only to add or subtract orders or families but also to verify the sensitivity of the groups. Such adaptation could be done at an 'eco-region' level, for example, all northern cold-water European countries are likely to have similar invertebrates, which will need to be separated from the African ones which will also be separate from the Asian ones. We are finding that the differences are relatively minor, however, and a single index that includes the orders and families most important for ALL regions of the world, is viable as a citizen science instrument. A generalized and indicative global index, using the miniSASS methodology, is thus possible for citizen scientists.

miniSASS has many strengths, as well as a number of challenges. These are tabulated in Table 1.

A summary of SDG indicator 6.3.2 methodology

SDG indicator 6.3.2 is reported by countries as the *proportion of bodies of water with good ambient water quality*. It is one of two indicators of Target 6.3 which aims to improve the quality of water in rivers, lakes and groundwaters by reducing pollution. Level one monitoring of the official indicator methodology (UN Water, 2018) relies on water quality data from *in situ* measurements and the analysis of samples collected from rivers, lakes and aquifers. Water quality is assessed by measuring physical and chemical parameters that reflect natural water quality, together with major human impacts on water quality. Level two reporting can include any type of water quality data that can be used to classify a body of water. Examples include data from citizen initiatives, Earth observation products, biological approaches or additional water quality parameters are not included in the core Level one list.

The indicator methodology stipulates that countries are divided into river basin-based reporting districts, which are further divided into smaller water body units. These small hydrologically defined units are classified based on the results of measurements of five core parameter groups: oxygen, salinity, nitrogen, phosphorus and acidification. These data are compared to numerical target values, and if a compliance rate of 80% or more is achieved, a water body is classified as having 'good' ambient water quality.

---

[4] We have been told that macro-invertebrates pre-date the dinosaurs and that their ancestors were present on the mega-continent, *Pangea*, before it split up (*Pers. Com.* Prof. Bruce Rubridge, Palaeontology, University of the Witwatersrand wits.ac.za). This offers an alternate explanation as to why one finds similar macro-invertebrates, which can be used as indicator species, in most perennial rivers of the world!

Table 1 | miniSASS strengths and challenges.

Strengths	Challenges
It supports the global trend to engage citizen science. This is especially relevant for realizing SDG 6 (b). Crowd-sourced data have its own unique validity; while each sample may not have a high level of confidence, the sheer number of data records strengthens the research rigor. The approach engages citizens with the SDG agenda, enabling them to contribute to a global effort. This could become a VERY big promotion for the SDGs. miniSASS data have been used to demonstrate citizen science participation in monitoring compliance with water quality objectives. The miniSASS data management system allows for the clustering of the data into a time-series. This means progress over time can be measured and compared at the same site. miniSASS costs very little to use. Simple apparatus such as a net, which can be home-made, and a reference sheet that is available as a free down-load, on the website, strengthens the miniSASS study. Since the macroinvertebrates are visible to the naked eye, the shape and form are of most importance. This means that advanced competence in languages such as English is not essential. Indeed, nine-year-old isiZulu speaking children are able to master the technique. miniSASS materials are also available in other languages such as French, Afrikaans and isiZulu.	Participants need to learn how to do miniSASS. Although there are simple instructions and tutorials on the website, participants learn best in the field with an experienced person. Citizen science by nature requires a level of coordination. At present, the coordination is provided by Ayanda Lepheana and *GroundTruth* with website support from SAIAB[a] and SAEON[b], both South African government-supported institutions. If the miniSASS data are to be used for SDG reporting globally, it will require additional support. Approval of data by the government may be challenging if the government prefers to 'be in charge' and not receptive to citizen science input and participation. To what extent will governments embrace and support the democratization of science? Each country will have the ability to either embrace citizen science to its fullest, or for government officials to use the method themselves as a low-tech monitoring method. While the latter is not invalid, it is unlikely that it will result in the high number of data that will be collected by a strong citizen science programme.

[a]South African Institute for Aquatic Biodiversity http://www.saiab.ac.za.
[b]South African Environmental Observation Network.

Case study: using miniSASS data to generate an SDG indicator 6.3.2 score

In many parts of the world, there are significant data gaps, both spatially and temporally, in the water quality record which cannot be filled using 'conventional monitoring programmes'. Citizen science projects are one of several approaches currently being explored to see whether they can play a significant role in filling these data gaps.

The miniSASS data generated by citizens do not fit the requirements of Level one reporting for SDG indicator 6.3.2 as they are based on a biological assessment, rather than physico-chemical data collection, and are not sampled at fixed monitoring locations. In this study, an indicator 6.3.2 score was generated that could be described as a 'Level two report' based on the current methodology (UN Water, 2018).

In order to calculate an indicator score, a data-rich area was selected as shown in Figure 1. The Department of Water and Sanitation of South Africa defines drainage basins for water resource management purposes. The case study area aligns with Drainage Basin U. River water bodies were defined using the HydroATLAS Level 10 units (Lehner & Grill, 2013) and a total of 129 river water body units were defined for the drainage basin. Each miniSASS data record was allocated to a water body unit based on spatial location, and the scores compared to the Fair/Good boundary of the ecological categories for the two river types ($\geq$5.9 for sandy-type rivers and $\geq$6.2 for rocky-type rivers). A water body was classified as 'good' if 80% of samples met this target.

In order to calculate the SDG indicator score, the proportion of water bodies within the basin with good water quality were calculated. Of the 129 river water bodies defined, 40 had miniSASS monitoring data and 6 of these were classified as 'good'. This thus yielded an SDG indicator score of 15 for this basin ($6/40*100 = 15$).

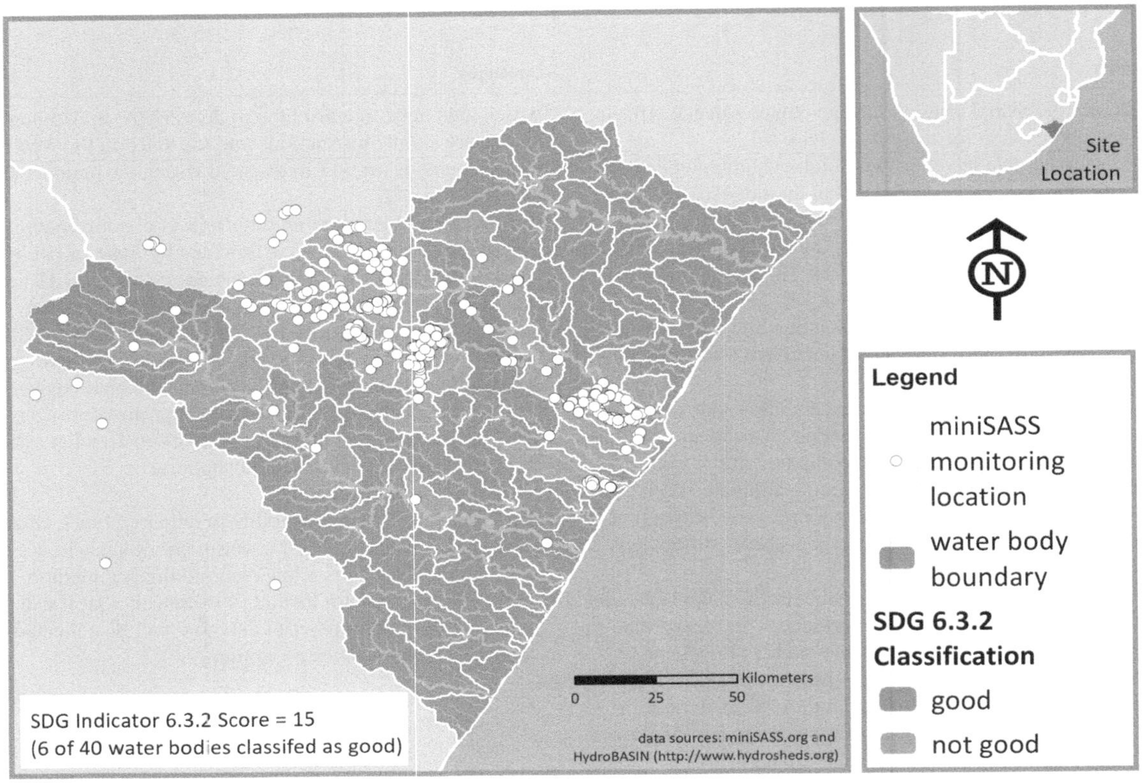

Fig. 1. | Example of miniSASS-generated SDG indicator 6.3.2 score for a single drainage area, in KwaZulu-Natal, South Africa.

One key finding of this analysis was that the monitoring effort differed considerably between water bodies. The most data records for any single water body were 183, whereas several had only a single record. This disparity in monitoring effort results in some water bodies being classified with a much higher degree of confidence compared to others. A further development of this method could include specifying a minimum data requirement to ensure that water bodies are classified equally and reliably. In this example, setting a minimum data requirement of five monitoring records per water body resulted in 17 water bodies being monitored, none of which were classified as 'good'. In addition to defining minimum data requirements, future testing could include optimizing monitoring network design to improve data collection coverage.

miniSASS as a proxy for formal water quality monitoring programmes

Much of the strength of miniSASS lies in its ability to generate large amounts of data through citizen science and crowdsourcing. Although these data may not always be entirely valid, due to the fact that many of the people who collect it may not have received formal training, it has its own unique validity as the sheer number of data records strengthens rigour (Holt, *et al.*, 2013). In order to assess the ability of miniSASS to act as a proxy for formal water quality monitoring programmes, miniSASS data were compared against data collected by the River Ecostatus Monitoring Programme (REMP). The REMP assesses the ecological condition of South Africa's rivers based on a rapid assessment of aquatic macroinvertebrates, using the Macroinvertebrate Response Assessment Index (MIRAI).

The 2017/2018 REMP data were compared against the miniSASS data collected for the same time period. In 2017 and 2018, a total of 522 miniSASS data entries (234 and 288 in each year, respectively) were recorded across South Africa. Of these, 13 were recorded within a 500 m radius of a REMP monitoring point and 98 within a 5 km radius (Figure 2). However, only 28 REMP sites within 5 km of miniSASS sampling points and 11 within 500 m of miniSASS sampling points had REMP data recorded during this time. Of the 11 miniSASS entries recorded within 500 m of REMP monitoring sites, four yielded the same ecological category as those reported at the REMP sites, three yielded an ecological category one category below those reported at the REMP sites, and the remaining four yielded ecological categories two categories below those reported at the REMP sites. The frequency of samples recorded in each ecological category by the REMP and miniSASS data collection is presented in Table 2.

Although the results assessed here do not provide conclusive evidence of the relationship between miniSASS and REMP data, they provide some indication that miniSASS is able to yield results similar to those derived from the REMP, particularly for rivers in a lower ecological condition (C, D and E). This, in combination with the fact that the method has been extensively assessed against SASS data and is strongly rooted in rigorous statistical evaluations, suggests that, where rigorous data collection may not be feasible due to resource constraints, miniSASS can provide defensible water quality results. Considering that many of South Africa's rivers are in a degraded state, a citizen science tool that can be used to provide evidence of a river's degraded state and draw attention to the plight of the freshwater resources, is of immense value.

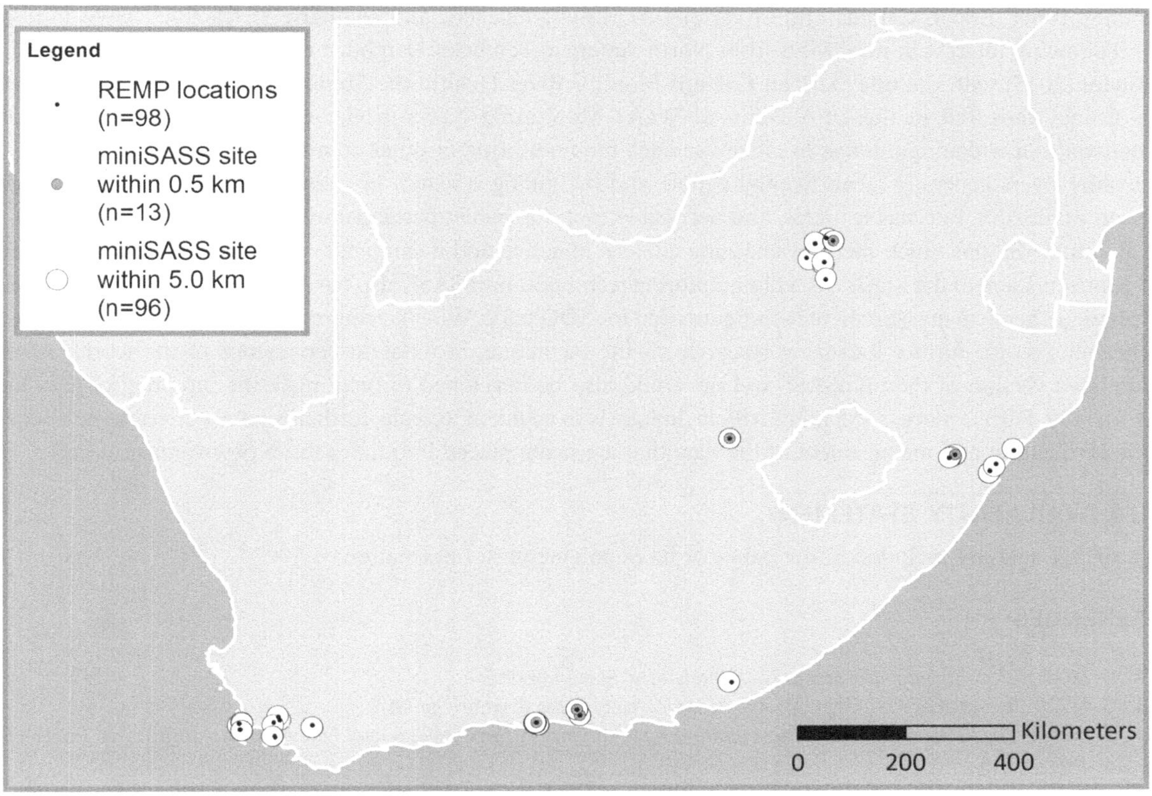

Fig. 2. | miniSASS sites in South Africa within a 500 m radius of REMP monitoring sites.

Table 2 | The frequency of samples in each ecological category recorded by miniSASS and REMP monitoring in 2017 and 2018.

Ecological category	miniSASS	REMP
A	0	0
B	0	0
C	1	7
D	5	4
E	5	0
Total	11	11

Only samples collected at miniSASS and REMP sampling points within 500 m of each other are displayed.

CONCLUSIONS

Human-created problems, that are compromising our water resources, require human-centred solutions. Simply communicating messages to people, through causal approaches, are unlikely to have any long-term, positive, effects. By involving people through co-engaged, enabling orientations such as citizen science and miniSASS, different, more inclusive and participatory, approaches become possible. This work is especially important if SDG 6 (b), stakeholder engagement, is to be realized.

The miniSASS biomonitoring technique has been successfully demonstrated in various parts of the world including India, Brazil, Canada and in various countries in eastern and southern Africa (refer www.minisass.org). Following interest in miniSASS from North America, Tembeka Dambuza developed an article (Dambuza & Taylor, 2015) with the title 'African Citizens Monitor River Health: the Stream Assessment Scoring System'. This article appeared in the USA National Water Monitoring News acwi.gov/monitoring *Sprint*, 2015 and demonstrates a widening interest in citizen science biomonitoring in other countries.

Healthy rivers generally mean healthy people and by linking research processes to indicator species such as through miniSASS, one has a useful, and accessible tool, for public participation (Graham, 2013; Taylor, *et al.*, 2013). miniSASS has much merit in engaging citizens in active and meaningful research that is real and applied and connects them to the SDGs. As a biomonitoring technique, miniSASS also has the potential for complementing the physico-chemical measurement techniques used for SDG 6.3.2. What is required, however, for greater universal applicability, is for further localizing research on the technique, in different eco-regions of the world. A more streamlined version of the miniSASS website could also be developed to help make the inputting of data more user-friendly. Furthermore, if the miniSASS technique is to be taken to scale, further resources need to be allocated to the verification and management of the data that are being placed into the website (www.minisass.org).

DATA AVAILABILITY STATEMENT

All relevant data are included in the paper or its Supplementary Information.

REFERENCES

Beck, U. (1992). *Risk Society: Towards a New Modernity.* Sage, London.

Beck, U. (1995). *Ecological Politics in an Age of Risk.* Polity Press, Cambridge, MA.

Beck, U. (1997). *The Reinvention of Politics: Rethinking Modernity in the Global Social Order.* Trans. Mark Ritter. Polity Press, Cambridge, MA.

Bhaskar, R., Frank, C., Hoyer, K. G., Naess, P. & Parker, J. (2010). *Interdisciplinarity and Climate Change: Transforming Knowledge and Practice for Our Global Future.* Routledge, London.

Bonney, R., Ballard, H. & Jordon, R. (2009). *Public Participation in Scientific Research: Defining the Field and Assessing its Potential for Informal Science Education.* CAISE, Washington, DC.

Dallas, H. F. & Day, J. A. (2004). *The Effect of Water Quality Variables on Aquatic Ecosystems, A Review.* Research Report No. 224/04, Water Research Commission (WRC), South Africa.

Dambuza, T. & Taylor, J. (2015) African citizens monitor river health: the stream assessment scoring system. *USA National Water Monitoring News.* acwi.gov/monitoring Sprint 2015.

Dickens, C. W. S. & Graham, P. M. (2002). The South African scoring system (SASS) version 5 rapid bioassessment method for rivers. *African Journal of Aquatic Science 27*, 1–10.

Fien, J. (2003). Education for a sustainable future: achievements and lessons from a decade of innovation, from Rio to Johannesburg. *International Review for Environmental Strategies 4*(1), 5–20.

Fritz, S., See, L., Carlson, T., Haklay, M., Oliver, J., Fraisl, D., Mondardini, R., Brocklehurst, M., Lea A. Shanley, L. A., Schade, S., When, U., Abrate, T., Anstee, J., Arnold, S., Billot, M., Campbell, J., Espey, J., Gold, M., Hager, G., He, S., Hepburn, L., Hsu, A., Long, D., Masó, J., McCallum, I., Muniafu, M., Moorthy, I., Obersteiner, M., Parker, A. J., Weisspflug, M. & West, S. (2019). Citizen science and the united nations sustainable development goals. *Nature Sustainability 2*, 922–930. Available at: www.nature.com/natsustain.

Graham, M. (2013). Species, healthy rivers and healthy people. *Environment 14*, 14–16.

Graham, M. & Taylor, J. (2018). *Development of Citizen Science Water Resource Monitoring Tools and Communities of Practice for South Africa, Africa and the World.* WRC Report No. TT 763/18, Pretoria.

Graham, M., Dickens, C. W. S. & Taylor, R. J. (2004). miniSASS – a novel technique for community participation in river health monitoring and management. *African Journal of Aquatic Science 29*(1), 25–36.

Holt, B. G., Rioja-Nieto, R., MacNeil, A., Lupton, J. & Rahbek, C. (2013). Comparing diversity data collected using a protocol designed for volunteers with results from a professional alternative. *Methods in Ecology and Evolution 4*, 383–392. British Ecological Society, London.

Jewitt, G. J., Sutherland, C., Browne, M., Stuart-Hill, S., Risko, S., Martell, P., Taylor, J. & Varghese, M. (2020). *Enhancing Water Security Through Restoration and Maintenance of Ecological Infrastructure: Global Lessons From the Umngeni Catchment, South Africa.* WRC Report No K5/2354. Final Report to the Water Research Commission.

Kemmis, S. & Mutton, R. (2012). Education for sustainability (EfS): practice and practice architectures. *Environmental Education Research 18*(2), 187–207.

Lehner, B. & Grill, G. (2013). Global river hydrography and network routing: baseline data and new approaches to study the world's large river systems. *Hydrological Processes 27*(15), 2171–2186. Available at: www.hydrosheds.org

O'Donoghue, R., Taylor, J. & Venter, V. (2018). How are learning and training environments transforming with education for sustainable development? In: *Issues and Trends in Education for Sustainable Development.* Leicht, A., Heiss, J. & Byun, W. (eds.). UNESCO, Paris. Available at: http://unesdoc.unesco.org/images/0026/002614/261445E.pdf

Pocock, M. J. O., Chapman, D. S., Sheppard, L. J. & Roy, H. E. (2014). *Choosing and Using Citizen Science: A Guide to When and How to Use Citizen Science to Monitor Biodiversity and the Environment.* Centre for Ecology & Hydrology, Natural Environment Research Council, Scottish Environment Protection Agency, Edinburgh.

Taylor, J. (2010). *Education for Sustainable Development: Perpetuating Myths or Bringing About Meaningful Change?* Global Environmental Research. United Nations University, Japan, pp. 187–192.

Taylor, J. (2014). Shaping the GAP: ideas for the UNESCO post-2014 ESD agenda. *Journal of Education for Sustainable Development 8*(2), 1–9. doi:10.1177/0973408214548369. SAGE Publications (Los Angeles, London, New Delhi, Singapore and Washington DC). Available at: www.sagepublications.com

Taylor, J., Graham, M., Gibixego, A. & Bruton, S. (2013). No rocket science – let the 'nunus' tell their story. *Stockholm Waterfront 3*, 14–15. November 2013. Available at: www.SIWI.ORG/WATERFRONT.

UN Water (2018). *Integrated Monitoring Guide: Step-by-Step Methodology for Monitoring Water Quality (6.3.2).* Available at: http://www.unwater.org/publications/step-step-methodology-monitoring-water-quality-6-3-2/.

Wals, A. E. J. (2007). *Social Learning Towards A Sustainable World: Principles, Perspectives and Praxis.* Wageningen & Gothenburg Universities, Wageningen.

First received 11 December 2020; accepted in revised form 16 August 2021. Available online 2 September 2021

© 2022 The Authors From Knowledge and Capacity Development to an Implementation Science: doi: 10.2166/wp.2022.210

Making the invisible, visible: 3D aquifer models as an effective tool for building water stewardship in Maharashtra, India

Eshwer Kale *, Marcella D'Souza and Sarita Chemburkar

WOTR Centre for Resilience Studies (W-CReS), Watershed Organisation Trust, The Forum, 2nd Floor, Padmavati Corner, Pune-Satara Road, Pune, Maharashtra 411009, India

*Corresponding author. E-mail: eshwer.kale@wotr.org.in

EK, 0000-0002-6956-5240

ABSTRACT

Frequent droughts, growing water demand, and the ease of accessing groundwater, particularly through boreholes, have led to an over-dependence on groundwater to meet notably agricultural demands in India. This has led the Watershed Organisation Trust (WOTR) to motivate and capacitate rural communities for 'water stewardship' in which the village as a whole comes together to learn about its resources to manage them better. However, improving the management of the water resource is challenging as the general perception is that the water found below one's farm belongs to the landowner. Moreover, what is below the surface is not visible, therefore not well understood. Thus, WOTR designed a water stewardship initiative (WSI) with the Community-Driven Visual Integrator (CoDriVE-VI or CDVI) as one of the important tools; it produces an operational 3D map of the local aquifer. WOTR implemented the WSI in villages in Maharashtra, India. This paper evaluates the impacts and potential of the WSI and CDVI. We observed that the visual effect of these 3D models of aquifers helps local communities and farmers better understand and appreciate groundwater as a shared resource. This tool is effective in promoting water stewardship, achieving behavioural change in water users, and helping implement groundwater laws and policies.

Key words: Aquifer management, Behavioural change, Groundwater governance, Resilience building, Water budget, Water stewardship

HIGHLIGHTS

- The behaviour of groundwater users is discussed and tools are presented to align it with sustainable aquifer use.
- There are too few pilots on aquifer management; most pilots are at the scale of the village or watershed.
- This pilot brings science, policy, and practice together and improves groundwater governance.
- Essential insight is gained to protect depleting aquifers through community engagement.

GRAPHICAL ABSTRACT

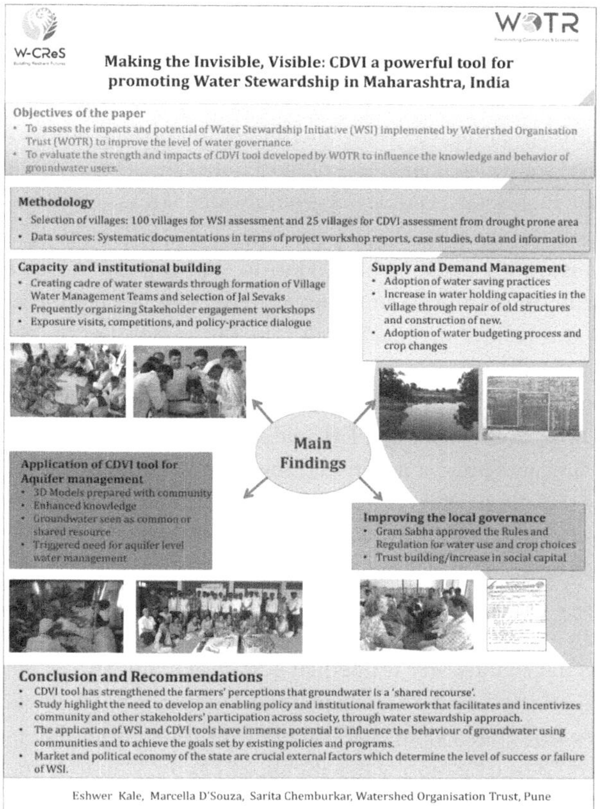

INTRODUCTION

Groundwater is a crucial component of water resources as it constitutes about two-thirds of the world's freshwater resources (Chilton, 1992). It is also broadly described as the world's 'hidden treasure': an invisible resource that provides safe drinking water and livelihood security to millions of people in arid regions (Chevalking *et al.*, 2008). India is the largest groundwater extracting country, and thus the groundwater is the lifeline for drinking, domestic, industrial, livelihood, and irrigation purposes (Shah, 2013).

Although a precious and life-saving resource, groundwater has for decades faced multiple challenges at different levels, particularly in its management and regulation. Many approaches have been designed and tested for decades. However, practice has shown that at a broader level, these are often not sustainable. The groundwater resource is challenging to manage because of its invisible nature and the belief that groundwater is a private resource, which leads to its over-use and the drastic decline in groundwater levels. In addition, changing precipitation patterns increases evapotranspiration rates caused by rising temperatures, and changes in the runoff cycle affect groundwater recharge rates (Kale & D'souza, 2019). Therefore, besides the further development of available resources, proper groundwater management requires attention to the judicious utilisation of the resource to ensure its long-term availability.

As the result of groundwater development and deregulation over the last four decades, around 84% of the total net addition to the irrigated area in India has come from groundwater use (Shah, 2013), and nearly 90% of the rural water supply is being met from groundwater sources (Prakash *et al.*, 2013). However, in 2004, 28% of India's blocks (district sub-division) were already showing alarmingly high levels of groundwater extraction (Planning Commission, 2011) which rose to 31% in 2013 (CGWB, 2013). The State of Maharashtra, where the Watershed Organisation Trust (WOTR) implemented the water stewardship initiative (WSI) and where the present study is located, has 52% of its area prone to drought, and most of the state has deep layers of hard impervious basalt rock called the Deccan Trap. These conditions render this belt highly susceptible to groundwater stress. Drinking water shortages are a recurring phenomenon in many parts of the state. In the summer of 2015, for the first time, the government was forced to run special water trains over a distance of 300 km to supply potable water to water-scarce areas of the Latur district (Kale, 2017). Furthermore, the Groundwater Survey and Development Agency (GSDA, 2019) reports that in Maharashtra, water levels have dropped from 1 to 3 m in 245 out of 353 *talukas* (townships) between 2014 and 2019.

Climate change and changing monsoon patterns are raising a second challenge. Regarding climate change, India is one of the largest and most vulnerable regions (Thow & Blois, 2008). The hydrological cycle is already being impacted (Dragoni & Sukhija, 2008). Maharashtra has witnessed a seven-fold increase in drought events and a six-fold rise in the frequency of flood events in 1970–2020 (Chaterjee, 2021). According to a recent Indian Council of Agricultural Research study, the state also has one of the highest vulnerabilities to extreme weather events (Adhav *et al.*, 2021). In addition, the rising average global temperature increases the rate of evaporation of surface water (TERI, 2014).

The perception of groundwater as a private resource puts farmers into a competitive mode as they abstract it through wells and boreholes (Kale, 2017). However, access to groundwater is unequal across Maharashtra as the land ownership determines it. Only a handful of resource-rich farmers have an adequate resource to tap. The local agronomic and political economy also determines water use to a great extent. For example, just three water-intensive crops – rice, wheat, and sugarcane – consume almost 80% of irrigation water (Shah, 2013). In Maharashtra, sugarcane is on 4% of cropped area but consumes 70% of irrigation water (Parth, 2020).

Rainwater stored in lakes and ponds can be easily seen and the amount of storage can be measured; however, when it comes to groundwater, people have no idea of the quantity available in aquifers, as this is not visible to the naked eye. Thus, the increasing demand for groundwater and its inefficient use have led to its unscientific exploitation. In addition, due to customary laws in India, groundwater is seen to be attached to the land property which results in the farmers' unrestricted exploitation of water below their land and even beyond (Iyer, 2003; Ballabh, 2008).

Hence, the challenge is to mobilise and sensitise the communities to improve the water governance by making the invisible groundwater resource visible. This necessarily involves building institutions and capacities of villagers, and an understanding of the aquifer systems, including basic hydrogeology, groundwater flows, groundwater depth, and sustainable aquifer yields. Therefore, it is vital to investigate groundwater systems and share the knowledge with its users. Without such a shared understanding, it will be difficult, even impossible, to envisage collective management of the groundwater resource. The behavioural change of groundwater water users and their capacity building – in terms of perception of groundwater ownership, water use, and crop management practices – are essential strategies for achieving judicious use of groundwater.

APPROACHES FOR BEHAVIOURAL CHANGE AND CAPACITY BUILDING

The individual's and community's behaviour regarding their water-use practices is at the centre of this paper. Groundwater management in India has a rich history. Along with numerous policy prescriptions by the government at the national and state levels, various agencies have launched initiatives under the broad framework of Community-Based Natural Resource Management (CBNRM). Many civil society organisations have piloted Participatory Groundwater Management (PGWM) and even the management of groundwater at the aquifer level. The government of India has initiated the National Aquifer Management Programme (NAQUIM) and formulated policies to promote groundwater management at the aquifer level. Although these initiatives are well intended and based on the mapping of aquifers, the challenge remains to make villagers more water-literate and demystify the invisible aquifer systems; the goal is behavioural change by building capacity. In the literature, the Theory of Planned Behaviour, the Social Cognitive Theory, the Institutional Analysis and Development Framework (IAD), and the Governance Dynamic Framework (GDF) are widely used frameworks to analyse the behaviour of individuals as well as of the community as a whole.

The Theory of Planned Behaviour explains that an individual's behaviour is dependent on one's intention to perform the behaviour, where the intention is determined by an individual's attitude (beliefs and values about the outcome of the behaviour) and subjective norms (beliefs about what other people think the person should do or general social pressure). In this theory, intention is the most crucial variable in predicting behaviour change, suggesting that behaviours are often linked with one's motivation (Sommer & Sigmaringen, 2011). On the other hand, the Social Cognitive Theory proposes that people are driven not by inner forces but by external or environmental factors (Bandura, 1999). Environmental factors represent situational influences and the environment in which behaviour is performed, while personal factors include instincts, drives, traits, and other individual motivational forces. The theory suggests that even when individuals have a strong sense of purpose, they may not perform the behaviour without incentive. The IAD framework developed by Ostrom (2011) helps understand how actors/stakeholders act in particular action situations and how the various rules and norms shape their interactions, behaviours, and dynamics. This framework has been tested and applied in a range of conditions to systematically analyse the structure of situations faced by individuals and determine how rules and attributes of the surrounding environment and local community affect these situations over time (Smajgl *et al.*, 2009). While this theory acknowledges that individuals may act rationally in simple situations, it posits that in complex circumstances, e.g. in common-pool resource management such as with groundwater, individuals lack complete knowledge and means to act in an economically rational manner (Polski & Ostrom, 1999).

The GDF developed by Subodh Wagle attempts to present a scheme of interactions between actors' behaviours and the relevant public policies and development interventions. The framework is based on the premise that the agencies' policies or programmes are meant to influence various actors involved in a societal issue to align with the policy or programme objectives. However, in practice, these actors behave according to their thinking and expectations and respond to the actual or expected behaviour of the other actors. Thus, such a situation creates misalignment in expected behaviour (in policies and laws) and behaviour in the practice of actors. The model explains that this mainly happens because the actors are not fully prepared to take appropriate actions. These key preparedness elements of the actor are awareness, vision, willingness, and abilities. The preparedness elements are partly influenced by some innate factors related to the actors while partly influenced by existing policies, agencies, and programmes (Kale & D'Souza, 2019).

The analysis of the above models and frameworks allowed WOTR to develop a Theory of Change for its WSI. It was found that after years of development interventions in the water sector by government agencies and NGOs, water security and governance have not yet improved to an expected level. Maharashtra has a long history of watershed development work with watershed development programmes, which increased water availability in the villages, resulting in an increase in irrigation, agricultural output, and economic prosperity. However, as surface water got fully used this growth could not be sustained triggering a race among farmers for extracting groundwater, resulting in water scarcity and in more crop failure (Kale & D'Souza, 2018). WOTR concluded, after situational analysis and interviews, that multiple factors are responsible for this. The strong focus on the harvesting of water combined with little attention to building villagers' capabilities to sustainably manage the harvested water was identified as the key factor; market incentives and climatic factors strongly contributed to the worsening of the situation. The Theory of Change proposes that, for changing the actors' behaviour, the following are required conditions: (a) adequate and scientific information and knowledge, (b) sense of ownership, (c) skills and confidence, (d) incentives, (e) convening of all relevant stakeholders to build a constructive dialogue, and (f) subjective norms, such as social pressure through the *Gram Sabha* (local council) (D'Souza *et al.*, 2019).

This paper reports on a central element in the effort to equip rural communities and farmers with knowledge, skills, and the means to address these challenges. The paper is based on data and experiences from 100 villages in Maharashtra where the WSI is implemented by WOTR, a premier organisation working in India in the field of water management and sustainable land management practices. The WSI aims to facilitate a cognitive and organisational shift by bringing science, governance, and policy together at the level of practice and community action. The approach adopted in the WSI is of 'co-production of knowledge and learning for behavioural and institutional change' towards building community knowledge and enhancing capacities to deal more effectively with local water-use practices and varying weather conditions. This paper discusses WSI impacts and presents the data, information, and learning experience with the application of a dedicated tool, the Community-Driven Visual Integrator (CDVI), designed to develop an understanding of the collective nature of the groundwater resource and achieve more consensus in the communities for the sustainable use of aquifers.

METHODS

This section presents first the methodology adopted for assessing the impacts of the WSI and CDVI, and data sources. Second, the methods are described for preparing the CDVI models.

Methods and data sources

The WSI was implemented in 100 villages in Maharashtra from 2015 to 2018. During this period, several stakeholder engagement events were conducted leading to water supply and demand-side interventions, institutional strengthening, and capacity building. The CDVI tool was applied in 25 villages that were especially groundwater-stressed, out of the 100 villages, along with the overall WSI activities, with the aim to enhance groundwater literacy and mobilise communities for aquifer management. At the formal closure, the project was evaluated to assess the impacts and lessons learnt. The paper utilises these data but also builds on the field observations of the authors during the implementation and evaluation.

Preparing CDVI aquifer models

The CDVI tool (see below) allows us to develop 3D maps with the involvement of communities so that their knowledge on the local natural resources can be integrated, mainly concerning the local hydrogeology. This scientific tool is designed to allow consideration of the different attributes of an individual's behaviour and how these attributes can be addressed to enable these actors to take appropriate actions towards more effective water management. Thus, the tool clarifies and simplifies the sub-surface characteristics of groundwater-using communities so that they can understand its complex nature. This understanding and the identified concerns are further combined to create the basis for bringing the community together to discuss and agree on the collective management of aquifers. Thus, the process consists of, first, the preparation of CDVI models of the aquifer together with communities, followed by a detailed stakeholder engagement through a series of workshops to mobilise them.

The surface and sub-surface models are prepared by applying the CDVI tool (Chemburkar & Kale, 2021). Managing groundwater effectively requires a comprehensive and consistent database and adequate knowledge of the aquifers and other relevant aspects. To develop the groundwater potential map and aquifer specifications, surveys using different tools and observations are conducted, including the application of RS-GIS, satellite-based image analysis, geological mapping, well inventory survey, and geophysical survey. Geological data collection and mapping are the primary components in delineating the aquifer. The information derived from geological mapping marks the formations with potential sub-surface storage, surficial features, and cross-cutting structures. The survey is conducted by applying the Schlumberger array of Vertical Electrical Sounding (VES) method to gain a detailed understanding of geophysical aspects. Later the data is interpreted using IGIS inverse slope software. This graph gives the thickness and resistivity of different geo-electric layers within the sub-surface. The detailed well inventory of selected wells helps to understand the local geology and groundwater yield in the selected geography. The groundwater potential map is obtained by synthesising the data and information generated by applying the above tools.

RESULTS AND DISCUSSION

Application of the CDVI model at the aquifer level

Watershed development has proven instrumental in convening people to manage their degraded landscapes. However, while water is harvested on the surface and from the aquifers, the shallow and deep aquifers generally do not follow the watershed boundaries. Since groundwater plays a major role in the livelihoods of rural households, it is essential to bring people together who share an aquifer for its appropriate management. Current Indian policies and laws also strongly advocate the management of the groundwater at the level of aquifers rather than village or watershed boundaries. For example, section 29.1 of the Maharashtra Groundwater Act 2009 states that 'The State authority shall constitute a Watershed Water Resources Committee (WWRC) in the notified areas comprising of an area of more than 11 villages, which share a common aquifer.' The WWRC is to comprise representatives of the participating villages, in addition to various experts at the *taluka* level, who have to collectively manage the aquifer sustainably and judiciously. Section 9.9 emphasises that mass awareness and training are essential for the WWRC members as well as for all stakeholders. A pilot on aquifer management was set up under the WSI along these lines in the Jalna district, Marathwada region. Applying the CDVI tool, an aquifer was identified and delineated which is shared by 14 villages in the Bhokardan block. This aquifer is known as the 'Malegaon aquifer' as it is the central village in the aquifer which is shared with the surrounding 13 villages. The main objectives behind the Malegaon pilot were generating awareness, sensitising and mobilising the villagers, building a cadre of 'Water Stewards' (*Jal Sevaks*, trained village youths) who understand their responsibility of managing the aquifer sustainably, and motivating the villagers to adopt self-regulation to take appropriate management measures. As an output of this process, Village Water Management Teams (VWMT) were formed in 14 villages, and *Jal Sevaks* facilitated VWMTs for planning and execution of water management practices. These 14 VWMTs were federated in the Aquifer Management Committee (AMC). Each *Gram Panchayat* (local government executive) nominated two members from its VWMT to the AMC.

Once the AMC was formed, a CDVI model was prepared for surface and sub-surface with the involvement of key villagers. Thereafter, AMC members took these models to each of the villages and presented them in village *Gram Sabhas* and village meetings. The Malegaon AMC eventually counted 15 members, one representative from each village, and the *Jal Sevak* on this committee. A critical indication that the Malegaon AMC is committed is that they have registered the AMC under the 1860 Societies Registration Act by contributions from members to pay the registration fees and for covering daily expenses.

The presentation of the maps to the 14 villages generated ample discussion as well as consensus on many points (Figure 1). Participants understood well that they all draw groundwater from the same aquifer, and thus need to prioritise more efficient groundwater use. Eight stakeholder engagement workshops were held with the AMC to develop their capacity for aquifer management. Although oftentimes challenging, participants recognised the precious value of the groundwater resource and agreed to take action to avoid further depletion. This led to (a) plans to harvest water to recharge their aquifer (a few structures have been constructed to address this), (b) demand management through appropriate crop planning, (c) water-saving by micro-irrigation, and (d) the formulation of village-level rules for water use and crop selection.

These events brought various stakeholders together for the first time. They helped participants to talk about their resources and needs and guided them to develop an objective understanding of their biophysical reality. It helped them become aware of issues related to their water resources, the availability and use of water, current agricultural practices, and how their decisions and current water-use patterns affect themselves, others, and future generations. Water stewards of different regions also reported that discussions and information received during the stakeholder engagement workshops have led to a new and deeper understanding of their water resources. The villagers reported that the stakeholder engagement workshops provided them with an opportunity to deliberate and discuss 'water' as a 'shared problem', leaving aside all other differences and dynamics of the village. They learned how to calculate the water budget and decide how to use water efficiently by taking collective actions. Notably, AMC members also highlighted the challenges they are facing in convincing the large irrigating farmers, mainly orchard owners, to follow the village water budgets and aquifer management plan. The orchard and cash crop cultivating farmers believe that for more income, they have to grow these crops as a conducive market exists. Thus, although the AMC is making progress in ensuring the efficient and sustainable use of aquifers, external forces such as a good market for water-intensive cash crops, changing rainfall patterns, and the behaviour of a few opportunist farmers remain important challenges. AMC members also shared that the strategy of social pressure by making rules in the *Gram Sabha* helped control opportunist farmers. Still, they also highlighted the need for effective regulation on water users by law and the district and block-level administration.

Application of the CDVI model at the village/watershed level

With the CDVI tool, CDVI models were prepared in 25 groundwater-stressed villages of the Ahmednagar and Jalna districts with the engagement of a team of key village informants representing older knowledgeable people, women, members of the *Gram Panchayat*, and other village institutions, adults, and youths. The joint CDVI model preparation proved an insightful journey for many. After preparing the model, many villagers reported that they could now visualise their entire village's topography as well as sub-surface aquifers, which they had never seen in a 3D format before. One of the participants from the Sangamner block stated 'I never thought that I could see our groundwater resource. The stakeholder engagement event made us realise that we pull out groundwater from the same aquifer. It is a common resource and therefore, important to manage together.' A *Sarpanch* (village head) from one of the villages from the Bhokardan block said 'attending the workshop enhanced our knowledge regarding the different rock types that are available in our village. It made us realise that if we do not manage water sustainably now, the coming generation will suffer. At the same time, women must be involved in these efforts as they face water problems in everyday life.'

WSI outputs and impact

As discussed earlier, the CDVI is one of WSI's central tools. Although the WSI effort has encountered difficulties and challenges (such as bringing irrigating farmers to the consensus and ensuring that they reduce their groundwater consumption by changing crops), villagers have progressed by coming together and initiating the change.

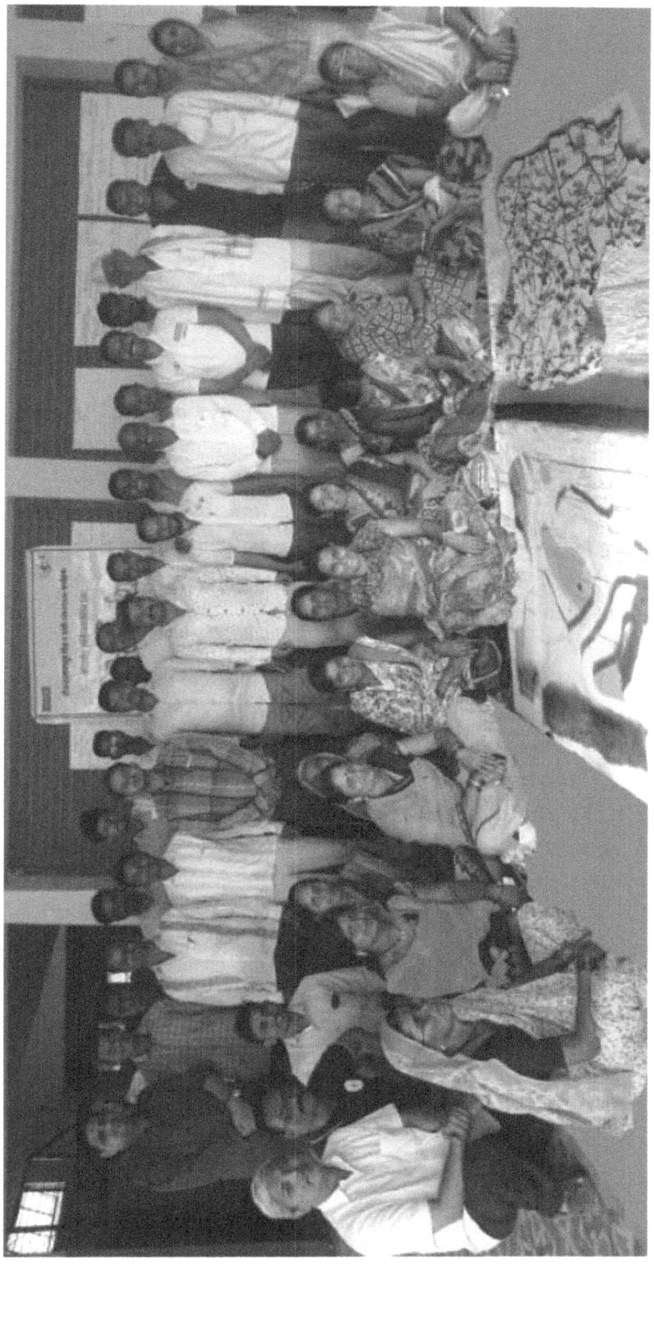

Fig. 1 | AMC members in a workshop at Bhokardan; the aquifer and watershed maps are placed on the floor.

Of the 100 villages where WSI work was carried out in Maharashtra, the performance of 46 villages eventually was considered satisfactory, 35 were moderately satisfactory, and 18 villages have underperformed (D'Souza *et al.*, 2019). The performance of these villages is assessed based on the following performance indicators: (a) the quality of water budgeting processes followed, (b) the rules agreed upon regarding crop practices and water use, and the supply and demand side of water management, and (c) the level of functioning of VWMT in these villages. Most members of VWMTs in these villages confirmed awareness of the causal relationship between the water crisis facing them and their own water usage and management practices. In all the villages, communities have drawn up water stewardship action plans (such as water budgeting and its follow-up in terms of supply and demand-side management), and 75 villages have submitted these plans to the government authorities to seek the convergence of different government schemes and programmes. Information about their villages' poor water status and the water demand encouraged the VWMTs and the *Gram Panchayats* to frame village-specific rules as guidance. These rules are discussed and then endorsed by the *Gram Sabha*. Seventy-eight villages formulated rules, such as a ban on drilling new boreholes, limits on the depth of boreholes, and other such rules as acceptable to them. The process of stakeholder engagement through the capacity-building workshops of the VWMTs was found highly effective to achieve these impacts.

Behavioural changes at home and in the fields were observed in many villagers. Around two thousand farmers were found to have adopted water-efficient technologies (such as micro-irrigation and mulching), and the adoption of better farming techniques has increased across all villages. Moreover, additional water harvesting capacity in all the villages has increased by about 9 billion litres through community contributions and governmental action. The VWMTs and *Jal Sevaks* played an important role in sensitising people and organising them to undertake the necessary actions at the village level. The stakeholder engagement events provided a platform for village representatives, experts, service providers, and government agencies. On the part of the government functionaries, the events helped them appreciate the underlying causes of the local water crisis and led to a significant buy-in which resulted in support to the committed measures in these villages. Equally important, it helped enhance the knowledge, skills, and confidence of villagers and the *Jal Sevaks* in their ability to address the challenges they face and in accessing publicly financed benefits and entitlements. Thus, the WSI has provided valuable lessons in understanding the complex relationships and compulsions that influence behaviours that determine water access and use at the ground level.

It remains a challenge, however, to sustain the motivation of *Jal Sevaks* and VWMTs after the withdrawal of WOTR from villages. For this, a mixed response arose in around 50 villages, where *Jal Sevaks* and VWMTs are preparing the water budget and its follow-up measures. Therefore, WOTR remained in touch with these villagers to at least keep encouraging them to continue their good work. Secondly, limitations were found also in achieving the equity goal as it proved challenging to gain the cooperation of the large and irrigating farmers for changing their practices; however, the social pressure through the *Gram Sabhas* has been able to address this albeit in a limited manner.

CONCLUSIONS

The WSI effectively applied VWMTs to mobilise villagers for measures such as water budgeting and plans for water-saving and harvesting more water. In the villages where the CDVI tool was introduced, participants were encouraged to actively participate in preparing the model. The CDVI models helped villagers to understand their groundwater resources and agree upon and scale up good practices in the sustainable use of groundwater management, thus changing their perceptions and behaviours on groundwater, resulting in systematic improvement in the groundwater governance at the local level. Importantly, the stakeholder engagement process helped participants to recognise that their local groundwater system is a finite resource and that individual efforts alone cannot solve their groundwater problem. Thus, villagers are given a platform and tool to agree on decisions for using groundwater more efficiently through collective efforts, including tools like water budgeting, setting rules and regulations within the village, and land management measures. One of the positive outcomes is that women's participation over the years has increased in the WSI; however, these efforts have to be continued to ensure that women's participation can be sustained and institutionalised.

The implementation experience of the WSI highlighted the need for an enabling policy and institutional framework that can facilitate and incentivise community and other stakeholders' participation. However, to achieve sustained impact, particularly given that external factors keep changing, guidance and monitoring support to the community are required from time to time and for years beyond the duration of the initiative implementation period. Perception and behaviour change is a long-term process, and the capacity-building efforts thus need to be planned and funded accordingly. Therefore, these efforts must be accompanied by sustained and large-scale, multi-format sensitisation campaigns, capacity building and skill up-grading of water users, governance strengthening measures, and the constitution of developmental and regulatory agencies. The establishment of a mechanism that enforces related policies and regulations for the common good in a transparent, fair, and consistent manner is necessary if the culture and practice of 'water stewardship' is to become engrained in village life. The Maharashtra Groundwater Act of 2009 is a first step in this direction, but needs to be implemented effectively. Other states in the country will eventually have to legislate and enforce similar policies to secure social stability, economic growth, and environmental integrity. The application of WSI and CDVI tools has substantial potential to influence the behaviour of groundwater-using communities and help achieve the goals set by existing policies and programmes such as the Maharashtra Groundwater Act of 2009, the National Project on Aquifer Management (NAQUIM), the Atal Bhujal Yojana, the SDG-6 ('Clean water and sanitation for all'), and international commitments for climate adaptation.

The study suggests that generating knowledge and information jointly with people, and making scientific information accessible to people through stakeholder engagement, have great potential. In addition, it helps develop social capital in terms of trust and social pressure through the *Gram Sabhas* which can help to influence the behaviour of water users. However, the need exists to simultaneously address the external forces such as the market incentives for water-intensive crops and the effective implementation of water-related laws and policies through effective regulation on the ground.

ACKNOWLEDGEMENT

This paper is developed based on lessons learned while implementing the WSI. We are thankful to our colleagues at WOTR and W-CReS, Ankita Yadav, Prashant Kalasakar, Arun Dahale, Santosh Chaudhari, Mahesh Shelake, Abhijeet Kavathekar, and Anil Hiwale, for their valuable support in organising the various stakeholder engagement workshops and motivating the villagers to actively participate in groundwater management.

DATA AVAILABILITY STATEMENT

All relevant data are included in the paper or its Supplementary Information.

REFERENCES

Adhav, C. A., Sendhil, R., Chandel, B. S., Bhandari, G., Ponnusamy, K. & Ram, H. (2021). *Socio-Economic Vulnerability to Climate Change – Index Development and Mapping for Districts in Maharashtra, India*. http://dx.doi.org/10.2139/ssrn.3854297.

Ballabh, V. (2008). *Governance of Water: Institutional Alternatives and Political Economy*. Sage Publications India, New Delhi.

Bandura, A. (1999). Social cognitive theory: an argentic perspective. *Asian Journal of Psychology 1999*(2), 21–24.

CGWB (2013). *Ground Water Information: Dhule District Maharashtra*. Ministry of Water Resources, Central Ground Water Board, Government of India. Available at: http://cgwb.gov.in/District_Profile/Maharashtra/Dhule.pdf.

Chaterjee, B. (2021). Seven-fold increase in drought events, six-fold rise in frequency of extreme flood events in Maharashtra over 50 years: study. *Hindustan Times*, accessed 12 January 2021.

Chemburkar, S. & Kale, E. (2021). *Making the Invisible, Visible: Manual for Preparing Co DriVE-Visual Integrator to Overlay Surface and Sub-Surface Characteristics for Sustainable Groundwater Management*. Watershed Organisation Trust, Pune.

Chevalking, S., Knoop, L. & Steenbergen, F. (2008). *Ideas for Groundwater Management*. MetaMeta and IUCN, Wageningen, The Netherlands.

Chilton, J. (1992). *Water Quality Assessments – A Guide to Use of Biota, Sediments, and Water in Environmental Monitoring*, 2nd edn. UNESCO, WHO and UNEP, Nairobi/Paris.

Dragoni, W. & Sukhija, B. (2008). *Climate Change and Groundwater*. Geological Society, London.

D'Souza, M., Kale, E. & Pinjan, H. (2019). *A Step Towards Quenching Rural India's Thirst. Experiences and Learnings from the Water Stewardship Initiative in Maharashtra*. Watershed Organisation Trust (WOTR), Pune.

GSDA (2019). *Report of Predicted Water Scarcity for Drinking Water for 2019–2020*. Groundwater Survey and Development Agency, Pune.

Iyer, R. (2003). *Water: Perspectives, Issues and Concerns*. Sage Publications India, New Delhi, India.

Kale, E. (2017). Problematic uses and practices of farm ponds in Maharashtra. *Economic & Political Weekly LII*(3), 20–22.

Kale, E. & D'Souza, M. (2018). Water stewardship in rainfed agrarian Maharashtra. In: *Water Conservation and Saving in Agriculture: Initiatives, Achievements, and Challenges in Maharashtra*. Kulkarni, S. (ed.). Water Resources Department, Government of Maharashtra, Pune.

Kale, E. & D'Souza, M. (2019). Building resilience to climate change: water stewardship in rainfed agrarian villages in Maharashtra, India. In *Handbook Climate Change Resilience*. Leal Filho, W. (ed.). Springer, Cham.

Ostrom, E. (2011). Background on the Institutional Analysis and Development frameworks. *The Policy Studies Journal* 39(1), 7–27.

Parth, M. N. (2020). *Maharashtra Sugarcane Is on 4% of Cropped Area But Consumes 70% of Irrigation Water*. Available at: https://scroll.in/article/974522/marathwada-is-close-to desertification and-yet-it-wont-give-up-farming-water-guzzling-sugarcane (accessed 12 January 2022).

Planning Commission (2011). *Twelfth Five-Year Plan (2012–2017): Faster, More Inclusive and Sustainable Growth*. Sage Publications India, New Delhi.

Polski, M. & Ostrom, E. (1999). *An Institutional Framework for Policy Analysis and Design*. Working Paper No. W98–27, The Vincent and Elinor Ostrom Workshop in Political Theory and Policy Analysis, Indiana University, Bloomington.

Prakash, A., Sharma, M. & Chourey, J. (2013). *Water in India: Situation and Prospects*. UNICEF, FAO, South Asia Consortium for Interdisciplinary Water Resources Studies, Hyderabad.

Shah, M. (2013). Water: towards a paradigm shift in the twelfth plan. *Economic Political Weekly XLVIII*(3).

Smajgl, A., Leitch, A. & Lynam, T. (2009). *Outback Institutions: An application of the Institutional Analysis and Development (IAD) Framework to Four Case Studies in Australia's Outback*. DKCRC Report 31, Desert Knowledge Cooperative Research Centre, Alice Springs.

Sommer, L. & Sigmaringen, A. (2011). The theory of planned behaviour and the impact of past behaviour. *International Business & Economics Research Journal 10*(1), 91–110.

TERI (2014). *Assessing Climate Change Vulnerability and Adaptation Strategies for Maharashtra: Maharashtra State Adaptation Action Plan on Climate Change (MSAAPC)*. The Energy and Resources Institute, New Delhi.

Thow, A. & Blois, M. (2008). *Climate Change and Human Vulnerability: Mapping Emerging Trends and Risk Hotspots for Humanitarian Actors*. Discussion Paper, Maplecro. Available at: https://reliefweb.int/sites/reliefweb.int/files/resources/E5549F80673D1E38C12574B0004B698F-care_mar2008.pdf.

First received 14 September 2021; accepted in revised form 9 February 2022. Available online 24 February 2022

© 2022 The Authors From Knowledge and Capacity Development to an Implementation Science: doi: 10.2166/wp.2021.255

Assessing the societal adoptability of participatory water management: an application of the Motivation and Ability (MOTA) framework

Md Shibly Sadik [a,*], Leon M. Hermans [b,c], Jaap Evers [b], Hong Quan Nguyen [d,e], Malik Fida A. Khan [a] and Sadiq Ahmed [f]

[a] Center for Environmental and Geographic Information Services (CEGIS), House 6, Road 23/C, Gulshan-1, Dhaka 1230, Bangladesh
[b] IHE Delft Institute for Water Education, Westvest 7, 2611 AX Delft, The Netherlands
[c] Faculty of Technology, Policy and Management, Delft University of Technology, Delft, The Netherlands
[d] Center of Water Management and Climate Change, Institute for Environment and Resources, Vietnam National University, Ho Chi Minh City 70000, Vietnam
[e] Institute for Circular Economy Development, Vietnam National University, 01 Marie Curie, VNU Campus, Quarter 6, Linh Trung, Thu Duc Dist., HCM City, Vietnam
[f] Policy Research Institute of Bangladesh (PRI), House 16, 4th Floor, Road 10/ A, Banani, Dhaka 1213, Bangladesh
*Corresponding author. E-mail: ssadik@cegisbd.com

MSS, 0000-0001-9205-4791

ABSTRACT

Exploring whether society is sufficiently equipped and motivated to adopt planned interventions is vital for modern plan development trajectories. The Motivation and Ability (MOTA) framework offers a tool to assess the societal adoptability of plans by exploring stakeholders' motivations and abilities. It was originally developed to assess plan implementation feasibility for structural measures of flood management in the Mekong Delta. Further development is necessary before applying the tool in other contexts and for other types of planning interventions. Institutional measures like participatory water management (PWM) have long been recognized as essential elements for water management, but have so far also remained out of the reach of conventional planning assessment tools such as cost–benefit or cost-effectiveness analyses. This research, therefore, aims at extending the MOTA tool in the context of PWM reforms in Bangladesh. It does this by, first, further detailing the MOTA components and identifying indicators for quantification and, second, an expert validation and application of this framework for coastal communities in Bangladesh. Our results suggest that the MOTA framework is capable of informing policymakers and implementing agencies about how to enhance the stakeholders' motivation and ability to ensure an enduring implementation of PWM reforms.

Key words: Bangladesh, Motivation and Ability framework, Participatory water management, Policy implementation, Strategic planning

HIGHLIGHTS

- The Motivation and Ability (MOTA) framework appeared recently to capture the societal and institutional dimensions in assessing the implementation feasibility of structural measures.
- This research further extends the MOTA framework and tests whether this can be applicable in case of assessing the implementation feasibility of soft measures like participatory water management.

GRAPHICAL ABSTRACT

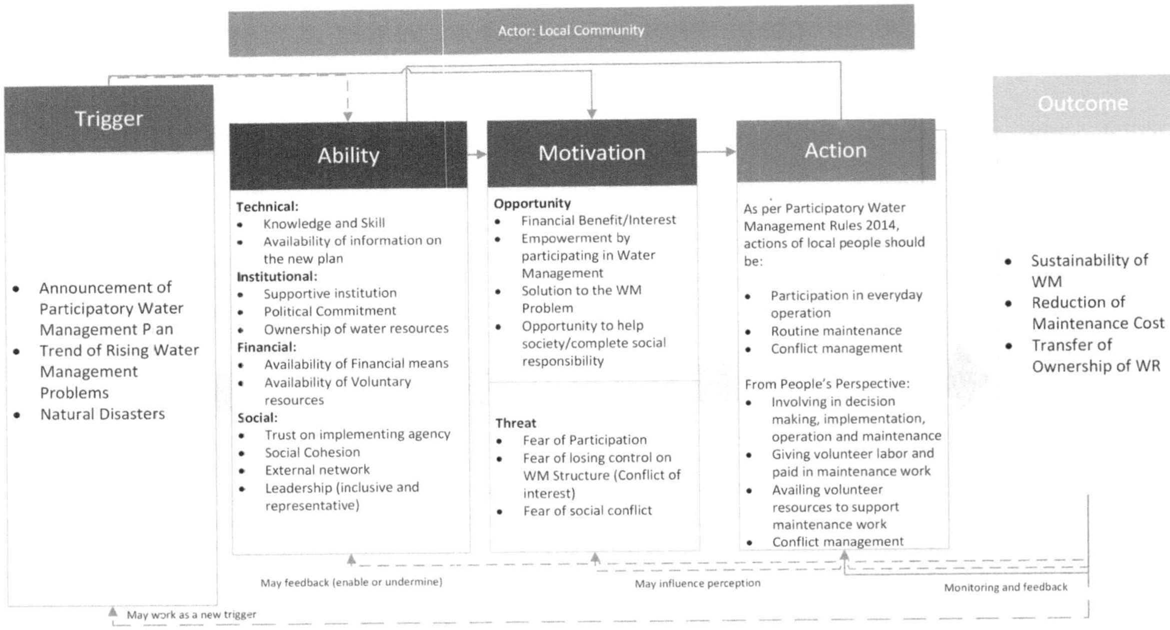

INTRODUCTION

The challenges in implementing strategic plans and policies are widely recognized: what is planned is not always implemented and what government agencies consider to be implemented is not always adopted by the society (e.g. Pritchett *et al.*, 2013; Phi *et al.*, 2015; Ménard *et al.*, 2018). This justifies why an assessment of whether the plan is implementable from the government's perspective and adoptable by society should be an integral part of the strategic planning and policy formulation.

The traditional planning assessment tools, e.g. multi-criteria analysis, cost–benefit analysis, environmental impact assessment, strategic environmental assessment, and robustness assessment, are mostly focused on the performance of a plan and its impacts on the economic and physical (e.g. water) system. These are required, but not enough to confirm that plan implementation by government agencies is feasible and that it can be successfully adopted by societal stakeholders (Nguyen *et al.*, 2019a). Phi *et al.* (2015) propose the Motivation and Ability (MOTA) framework (Phi *et al.*, 2015; Nguyen *et al.*, 2019a) to assess plan implementation feasibility, as dependent on governmental implementability (of the government and semipublic agencies involved in plan implementation and delivery) and societal adoptability of the target group (societal actors, such communities and businesses). This is done by taking into account the motivation and ability of the actors, and the opportunity and threats perceived by them, building on similar earlier methods in this area (Ajzen, 2002; Fogg, 2009; Hermans & Thissen, 2009). The tool is mostly applied to assess plans for capital-intensive large infrastructural strategies like building urban flood prevention infrastructures (Phi *et al.*, 2015) and more generic livelihood transformations – without further detailing of how to bring them about (Nguyen *et al.*, 2019b). While evaluating the implementation of strategic delta plans or water management policy, focusing on infrastructural measures or abstract transformation pathways will only cover parts of the plan. Institutional and governance-related measures[1] are considered as key components of long-term strategic plans, with critical relevance for plan implementation and sustainability (Brandes & Kriwoken, 2006; Korbee *et al.*, 2019).

The government of Bangladesh has recently formulated its first-ever long-term strategic plan, the Bangladesh Delta Plan 2100 (BDP 2100). It includes a diverse set of institutional measures (GED, 2018). Reform of participatory water management (PWM) is one of them. Implementation of such a strategic measure is influenced by institutional setting. Similar to infrastructural measures, it is essential to diagnose whether these institutional measures are feasible to implement. Most of the traditional planning assessment tools are not well suited to deal with the impacts of institutional measures and cannot be used to determine whether the society is capable and motivated to adopt such institutional measures. Therefore, the new challenge is how to diagnose the implementation feasibility of institutional measures as prominent parts in many long-term strategic plans. The earlier application of MOTA suggested the potential for including institutional measures (Korbee *et al.*, 2019; Nguyen *et al.*, 2019b), but it needs further development for contextualization and quantification. In this background, this research aims to further detail the MOTA framework and extend it to quantify societal adoptability in the context of an institutional measure. It does so for PWM reforms in Bangladesh. In this way, the research also aims to inform and aid the implementation of the BDP 2100 by bringing out detailed insights into the societal adoptability of proposed PWM reforms.

To that end, this paper first briefly discusses the theoretical background of the MOTA framework and the societal and institutional context of PWM. Thereafter, it explains the methods for the indicator selection, MOTA operationalization and scoring, and the indicator framework validation and test application. After that, it describes the main results: (i) an extended MOTA indicator framework and (ii) the outcome of the expert validation and test application. Finally, the results are further discussed with the outcomes and their implications, before concluding the paper with the key highlights of our work.

THE MOTA FRAMEWORK FOR ASSESSING STRATEGIC INSTITUTIONAL INTERVENTIONS

Earlier development of MOTA

The MOTA framework is essentially an actor analysis tool, which provides a measure to assess the likelihood of an action of an actor by assessing its motivation and abilities (Phi *et al.*, 2015). Building on existing actor analysis methods (Hermans & Thissen, 2009), the MOTA framework was designed to develop a more specific and operational instrument for quantitative stakeholder analysis in the water sector. Phi *et al.* (2015) developed the framework interlinking existing actor analysis methods and behavioral models (Ajzen, 2002; Fogg, 2009). Fogg's behavioral model is the base of the MOTA framework, which states that a behavioral action happens when motivation, ability, and trigger act together (Fogg, 2009).

The MOTA framework was originally developed to assess the implementation feasibility of a drainage improvement plan of Ho Chi Minh City. The implementation feasibility was defined by the adoptability of the plan by the

[1] We recognize that in the water resources management domain, often a distinction is made in hard and soft measures (Sovacool, 2011; Wesselink, 2016), in which the first relates to infrastructural measures and the latter to other policy instruments to influence target group behaviour and to manage our water resources. Yet, other fields of environmental management regard hard measures as regulatory coercive instruments and soft measures as incentivizing and stimulating measures, like a.o. knowledge dissemnination, voluntary agreements, and subsidies (Hertin *et al.*, 2004). In this paper, we considered the first explanation, yet we are reminded that "soft measures" are a problematic term, as they are often hard to implement. Therefore, we decided to move forward in this paper with infrastructural measures (measures that require construction in the physical environment, like dikes, weirs, and sluices) and institutional measures (measures that focus on the rules of the game and its players). Participatory management is thus an institutional measure as it influences the rules of the game, the institutionalized involvement of farmers in water resources management.

adopting actors (i.e. the society) and the implementability of the implementing actors (i.e. the government agencies involved in direct implementation) (Phi *et al.*, 2015). The theoretical framework was later implemented in the Mekong Delta of Vietnam to assess the societal adoptability of different livelihood transformation strategies of the Mekong Delta Plan (Nguyen *et al.*, 2019b). Thereafter, as the development of MOTA was continued, it traveled beyond the Mekong Delta and was applied in Bangladesh to anticipate farmers' climate change adaptation responses where the author adopted the MOTA framework in combination with a framed scenario approach (Kulsum, 2020).

Need for extending MOTA for the quantification of institutional measures

In earlier applications of MOTA in Vietnam, MOTA scores were subjectively quantified by experts and through surveys among stakeholders to compare the implementation feasibility of different alternative plans and strategies. In those quantifications, MOTA scores were constructed by further operationalization of MOTA elements for ability and motivation. Ability was quantified by a subjective measurement of its components, i.e. technical ability, financial ability, institutional ability, and social ability, following either a numerical scale (Phi *et al.*, 2015) or a custom-built Likert scale (Nguyen *et al.*, 2019b). In such quantification, components of ability remained relatively abstract and subjective. Phi *et al.* (2015) do point out this limitation in their original article and discuss the need for detailing out the MOTA components and modification of the quantitative techniques.

Later, MOTA applications for the Vietnamese Mekong Delta Plan provided more elaborated and transparent quantification processes (Nguyen *et al.*, 2019a, 2019b). Although MOTA was mostly applied in the cases of infrastructural measures, some applications were found assessing non-infrastructural measures in long-term plans, like agriculture transformations (Nguyen *et al.*, 2019b). On that note, we assume that the MOTA framework can be further adjusted and extended to address the implementation feasibility of institutional measures, which is needed in the BDP 2100 (Bangladesh Water Management Rules 2018 as well), i.e. PWM reform.

PWM and the need for exploring societal adoptability

In the scientific literature, with its origin in the political upheavals of democracy during the 1970s (Razzaque, 2009), the notion participation meant to be the participation of the governed in the government, and it was then defined from the perspective of the citizen power and democracy (Arnstein, 1969). Public participation, which is referenced as any form of engagement as minimum as information sharing or as highest as citizen control (Arnstein, 1969; Swapan, 2016), is not a new idea. Early forms of public participation were practiced in managing rural water resources in Europe around 1,000 years ago (Razzaque, 2009). Even in a developing country like Bangladesh, village-level flood management and irrigation infrastructures were traditionally built by the local community, during and before the 1900s (van Staveren *et al.*, 2017).

Contemporary forms of participation started appearing globally in planning and management due to the influence of international policies and rules, e.g. Stockholm Declaration 1972, Rio Declaration 1992, Agenda 21, Arhus Convention of 1999, Hague Declaration of 2001, Water Framework Directive of EU in 2000, and Berlin Rules of 2004 (Razzaque, 2009). With time and gradual development, participation as a theory of citizen rights and empowerment evolved to become an approach or tool of resource management and governance (Razzaque, 2009). It is also framed by different terminologies like public participation, popular participation, collective action, community-based/driven action, and stakeholder engagement with a core concept of people's involvement for a common objective (Kyamusugulwa, 2013). It is thus commonly defined as the inclusion of non-state actors in decision-making and implementation (Wesselink *et al.*, 2011; Mancilla García & Bodin, 2019).

In the case of water management where a participatory approach is recommended to deal with increasing water problems, future challenges, and uncertainties, mapping and analyzing stakeholders are more important and should be scientific (Stanghellini, 2010). Successful or effective implementation of PWM depends on the competent design and contextualization of participatory processes, within the local societal and political settings (Von Korff *et al.*, 2012). In addition, the capacity and willingness of state actors and the non-state actors alike are key determinants of the effectiveness (Tosun, 2000; Webler & Tuler, 2001; Kapiriri *et al.*, 2003; Hobbs & White, 2012; Von Korff *et al.*, 2012; Swapan, 2016).

Another importance of exploring willingness, motivation, and resources to participate is that participants might have different problems and different rationales than what are perceived by the implementing agencies (Wesselink *et al.*, 2011). In rural communities, PWM is implemented with the aim of transforming the water governance from state-controlled to community-controlled structures and thereby transfers the state's responsibility of managing water resources to the community (Sultana, 2009). Therefore, it is important to know the ability and motivation of the community to take such responsibilities. Public participation in developing countries is commonly criticized due to poor outcome, inefficacy, and lack of motivation (Tosun, 2000; Hobbs & White, 2012; Swapan, 2016). There is an argument that some basic physiological requirements must be met before motivating the community to participate and unfortunately communities in developing countries are socioeconomically deprived to meet such requirements which ultimately demotivate them to participate (Hobbs & White, 2012). The exploration and quantification of the communities' ability and motivation would tell whether the PWM plan will be adoptable.

PWM in Bangladesh and outlining what needs to be evaluated by MOTA

An important step of MOTA is to define the expected actions of the actors for which motivation and abilities are to be assessed. In this application of MOTA, the expected action of actors (local people in this case) is participation in water management. This participation has both normative perspectives set out by international literature and the legal and administrative perspective outlined in national rules and strategic plans, e.g. Bangladesh Water Management Rules 2014 and BDP 2100.

In Bangladesh, participation is defined differently in different policies and legal documents. The National Water Policy 1999 defined participation as a right of the people. The Guideline for Participatory Water Management 2000 (GPWM) defined it as a process of influencing of water management-centric decision. The Participatory Water Management Rules 2014 defined participation as a responsibility of the local people and it transferred the ownership of water management projects to water user groups. The ownership was defined as a package that includes the responsibility of everyday operation, regular maintenance, periodic maintenance, and emergency maintenance of medium- and small-scale projects without any financial arrangements. In practice, public participation is added to a project in compliance with a legal requirement to get the project approved by the Government. In the recently established BDP 2100, participation is again considered an important element for future sustainable water management. Based on the PWM Rules 2014 and the BDP 2100, we, therefore, considered that the expected participation of the local people in water management would include (i) participation in the planning and implementation process, (ii) taking responsibility for everyday operation, and (iii) taking responsibility of regular maintenance.

METHODS: EXTENDING THE MOTA FRAMEWORK FOR PWM IN BANGLADESH

Developing the indicator framework

This research is an integrative literature review. The literature was searched in scientific databases using relevant keywords like 'Participation', 'Participatory Water Management', 'Willingness to Participate', 'Evaluation of Participation', and 'Indicators of Participation'. The literature found was then selected purposively considering relevancy to the topic. A preliminary list of the relevant indicator was identified, reviewing indicators of willingness to participate (Cavalcanti *et al.*, 2010; Franzén *et al.*, 2015; Jennewein & Jones, 2016), adapting community participatory models (Swapan, 2016), cultural barriers to participate (Macnaghten & Jacobs, 1997), limits to community participation (Tosun, 2000), priority setting for participation (Kapiriri *et al.*, 2003), outcomes of participation (Rowe & Frewer, 2000; Perkins, 2011; Pineda *et al.*, 2014; Ballester & Mott Lacroix, 2016), good participation process (Krueger *et al.*, 2001; Dyer *et al.*, 2014), the success of participation (Özerol & Newig, 2008), and a few more relevant literature found by snowballing.

Thereafter, an exercise of indicator mapping (Supplementary Material, Figure 1) was performed to see interlinkages among the indicators and their relationships with MOTA components. This exercise helped to create a conceptual map of the indicator framework, resolve double counting, and exclude indicators irrelevant to the MOTA components. Besides, informal discussion with relevant experts (mostly from implementing agencies – the deputy team leader (DTL) of BDP 2100, DTL of the Blue Gold Program of Bangladesh Water Development Board (BWDB) implementing PWM, Chief Water Management of BWDB, and two more field level officers of BWDB implementing PWM) helped to refine the indicator list.

Once the list of the indicators was ready, each indicator was translated into a simple question to aid the quantification process during surveys or focus group discussions with local water users. Answers were scored on a numeric scale ranging from '0' to '5' (Supplementary Material, Table 1).

Expert validation and test application

The indicator framework was validated and tested by case application in a session during an interactive expert workshop. The expert workshop was a part of a 2-day MOTA workshop organized in Khulna, Bangladesh, in September 2020. The 30 workshop participants consisted mainly of staff from local non-governmental organization (NGOs), local relevant government agencies, local universities, and a national-level research institute, as well as a few graduate students from local universities. For the expert validation and application, the participants were grouped into seven groups (small farmer, medium farmer, large farmer, agricultural labor, shrimp farmer, rice farmer, and local civil society organization (CSO) and NGOs). The participants were asked to apply the MOTA indicator framework to quantify the societal adoptability of PWM in the south western coastal region of Bangladesh from the perspective of the actor groups they were assigned to. Each group first quantified the MOTA-ability indicators by making use of the indicator quantification questionnaire and weighted each indicator thereafter using a scale ranging from '0' (not important) to '5' (very important). The MOTA score for each ability component was then calculated as the weighted average. Each group presented their results to other groups. After the discussion, the groups were allowed to revise their indicator if judged necessary. Taking the final scores from the interactive expert group exercise, the final MOTA-ability score was calculated by averaging the score of the four components for each actor group, which was then normalized following the min–max method (OECD, 2008) and converted into a percentage score.

RESULTS: MOTA INDICATORS FOR THE SOCIETAL ADOPTABILITY OF PWM IN BANGLADESH

Triggers

In the original article on the MOTA framework (Phi *et al.*, 2015), the 'trigger' was defined as an event or a trend that can change the behavior of an actor by influencing his/her ability and perceived opportunity or threat. Since a trigger is a causative factor, the influence of a trigger would depend on the intensity of the trigger (Nguyen *et al.*, 2019a). However, for our case, the main triggers identified were the announcement of new plans for PWM, increasing pressures from water management problems, and natural disasters. Although these triggers could be quantified, we have not considered them for separate quantification because the measurement of ability, opportunity, and threat would be inclusive of the effects of the triggers.

Table 1 | MOTA indicators for quantifying societal adoptability.

Indicator	Justification
MOTA element: ability	
Ability Component 1: financial	
Availability of fund	Participatory activities from people's perspective need additional financing (Özerol & Newig, 2008). Case studies in Bangladesh found the unavailability of financial resources as an important reason for nonfunctional water user groups, which were established for ensuring community participation in a polder (Pineda *et al.*, 2013, 2014).
Availability of voluntary resources	Community participates in water management by giving voluntary labor or donating construction materials and such volunteer resources are considered important keys to the functionality of PWM in Bangladesh (Pineda *et al.*, 2014).
Ability Component 2: institutional	
Supportive institution	Legitimate assurance of reflection of the outcome in the policy or decision motivates people to participate (Carter & Hill, 2007; Perkins, 2011). The institutional framework to implement the participatory process must be designed to make the state actors remain actively involved in the participatory process (Jiménez *et al.*, 2019). Decentralization of the governance, ensuring an enabling environment for participation, makes the institutional framework supportive (Jiménez *et al.*, 2019).
Political commitment	Political commitment to ensure participation, the role of people in decision-making helps to achieve community trust in the participatory process and thereby motivate them (Cavalcanti *et al.*, 2010). For a developing country like Bangladesh where legislation is inadequate or lacks enforcement, strong political commitment from local leaders, local administrators, and implementing agencies might motivate people to participate.
Feeling of ownership	The feeling of ownership also equips the community to continue participation (Dyer *et al.*, 2014; Samaddar *et al.*, 2017). In PWM, the ownership implies that the community feels the water resources and the allied infrastructures are community resources, and therefore the operation and management of the project is the responsibility of the community. Such a feeling of ownership motivates the community to participate.
Ability Component 3: social	
Trust on state actor	Lack of trust in state actors, including the government, undermines their efforts of ensuring public participation (Macnaghten & Jacobs, 1997). Visibility or transparency of the participatory process and past examples of successful participation influence local people whether to trust the state actors (Rowe & Frewer, 2000; Samaddar *et al.*, 2015; Ballester & Mott Lacroix, 2016). If we take the indicator – 'trust on state actor', it also captures the influence of a good participatory process which ensures equity and visible role of each participant.
Social cohesion	Social cohesion is defined as the extent of connectedness and solidarity in society (Manca, 2014). Receiving cooperation from others and the internal relation of the society are most factors that influence willingness to participate (Cavalcanti *et al.*, 2010). Stronger social cohesion leads to better community participation (Desbureaux, 2018).

(*Continued.*)

Table 1 | Continued.

Indicator	Justification
External network	Most of the successful cases of participatory management were directly benefited by external agencies like NGOs and development partners (Samaddar *et al.*, 2017; Jiménez *et al.*, 2019). PWM cases in Bangladesh also support the statement that external supports help the mobilization of community and thereby motivate them to participate (Dewan *et al.*, 2014; Heering, 2014).
Inclusive and representative leadership	The opinion of a leader influences the community to accept any management measures (Cavalcanti *et al.*, 2010). Responsible (Lammerink *et al.*, 1999; Webler & Tuler, 2001) and committed leadership (Lammerink *et al.*, 1999) motivate people to participate and thus become an important condition of the effective participatory process (Krueger *et al.*, 2001). An inclusive leadership (Krueger *et al.*, 2001) protects the marginal from being excluded by the power elites and thus motivates local people to be in the participation process ensuring equity and free from elite capture. A representative leader is a link between the community and the external stakeholders (van Maasakkers *et al.*, 2014).
Ability Component 4: technical	
Knowledge and skill (on water management and participation)	Lack of technical knowledge often makes people shy to participate due to the fear that their opinion might be rejected by the technical people (Bardhan, 2000; Kapiriri *et al.*, 2003).
Availability of information on new plan	Announcement of a new plan acts as a trigger to perceive the community whether the new plan would bring opportunity or threat to them (Phi *et al.*, 2015). It also influences the ability by availing new information.
MOTA element: motivation	
Motivation Component 1: opportunity	
Financial benefit	Sometimes people due to their socioeconomic necessity expect financial benefits for participation in the planning process (Kapiriri *et al.*, 2003). In Bangladesh, financial benefits like the opportunity of earning money as labor also motivate people to participate in water management (*Source*: Interactive Expert Group Discussion). Moreover, people who could make productive use of water, e.g. farmers and fishermen, show more interest in participation in water management due to their financial benefits (Buisson *et al.*, 2017).
Resolving water management problem	Improving the management of common pool resources is a common motivation behind participatory management (Jennewein & Jones, 2016; Swapan, 2016). Therefore, a solution to the problem can be foreseen as an opportunity to participate.
Empowerment	Participation must bring an opportunity of being empowered by ensuring a visible role in decision-making (Krueger *et al.*, 2001; Ballester & Mott Lacroix, 2016). If people do not see this opportunity, they will not participate.
Fulfilling the moral and social obligation	When the community feels that the management of the local water resources is the responsibility of the community, they feel participation in water management is a moral thing to do to fulfill a social obligation. Such feeling appears in the community when they are aware of their ownership and right.
Motivation Component 2: threat	
Fear and shy to participate	In most of the cases, marginal people fear to participate (Tosun, 2000; Kapiriri *et al.*, 2003). The fear is that their objection or opinion can be used against

(*Continued.*)

Table 1 | Continued.

Indicator	Justification
	them (Tosun, 2000). Marginal people often feel shy to participate or to make any opinion, assuming that their opinion might be rejected by the technical experts (Kapiriri *et al.*, 2003).
Fear of social conflict	In a heterogeneous society, different groups of people might have different interests in water resources management. In such cases, people may think public participation in managing water resources, which may increase conflict (*Source*: experts' opinion during MOTA workshop in Khulna). On that argument, they refrain from participation.
Fear of losing control	It has been known during the MOTA workshop that people, especially large shrimp farmers who are controlling the water management infrastructures, oppose PWM.

Ability

Phi *et al.* (2015) defined the ability with three categories – financial, institutional, and technical. However, this extension of MOTA brings an important change by adding social capacity to the ability component of MOTA. Since social capital plays an important role in community capacity and drives the community toward a collective action (Krishna, 2002; Nakagawa & Shaw, 2004; Marré & Weber, 2010), for assessing societal adoptability, 'social capacity' has been added to the ability component of MOTA. The social capacity here refers to the social capital concept and includes social bonding and cohesion, trust, external network, and trust in state actors (Table 1).

Similar to the original MOTA framework, financial capacity refers to the availability of funds and budget. This research proposes the availability of funds and the availability of voluntary resources as two indicators of the financial capacity of the society to adopt PWM (Table 1).

The institutional capacity also follows the original MOTA framework, which is defined as institutional arrangements, coordination, rules, and regulation. To quantify the institutional ability, we have proposed three indicators – supportive institution, political commitment, and feeling of ownership (Table 1).

The technical capacities have also been defined as the original MOTA framework by referring to knowledge, skill, and information. For quantifying the technical capacity of the society to adopt the PWM plan, this research proposes two indicators – knowledge and availability of information on the new plan.

Motivation

Motivation is one of the important elements of MOTA (Phi *et al.*, 2015; Nguyen *et al.*, 2019a). In this extension of MOTA, motivation is defined as a degree of recognizing opportunity or threat from the proposed PWM reforms, which would drive water users in a community to decide whether to participate in water resources management. It thus adopts the reasoning of Phi *et al.* (2015) that the motivation is a function of perceived opportunity and threat.

Opportunity and threats

While extending the MOTA in the context of societal adoptability of PWM, we explored opportunities (advantages) and threats (disadvantages) of participation in local water management. For quantifying the opportunity of PWM, which would motivate the community to adopt the PWM plan, this research proposes four indicators (Table 1). On the other hand, three indicators have been identified to quantify the threat of the PWM, which could be perceived by the community and ultimately undermine the motivation. Those indicators are fear and/or shy to participate, fear of social conflict, and fear of losing control.

MOTA indicator framework for participatory water management

Based on the above discussion and the identified indicators in Table 1, an analytical framework of MOTA has been constructed to explain the societal adoptability of PWM reforms in Bangladesh (Figure 1). The extended MOTA framework (Figure 1) here shows the aforementioned triggers, ability, and motivation variables that influence the actions of water users in a local community concerning proposed PWM reforms. The expected actions of PWM have already been devised in the Participatory Water Management Rules 2014. However, such actions only reflect the perspective of the implementing agencies. While testing and updating the framework during the MOTA workshop, what would be the expected actions of PWM from people's perspectives were also identified. Finally, the framework also indicates the expected outcomes of PWM. The outcome again can act as a trigger and may feedback the ability as well. An outcome may also change the community's perception of opportunity and threat.

RESULT OF THE TEST APPLICATION OF MOTA INDICATOR FRAMEWORK FOR ASSESSING THE ABILITY OF COASTAL COMMUNITY IN BANGLADESH TO ADOPT PWM

Scores quantifying MOTA-ability indicators obtained by the interactive expert workshop are presented in Table 2, and a further detail (weightage of indicators and calculated MOTA-ability scores) is provided in Supplementary Material, Table 2. In addition to the table that explains that values of ability indicators are different for different livelihood groups, scores of different MOTA-ability components of different livelihood groups are shown in Figure 2, which illustrates the ability with reference to four ability components – financial, institutional, social, and technical.

In general, the ability scores of four components are low, below '3' except the CSO-NGO group which is practical because CSO-NGOs are institutes whereas others are livelihood groups (Figure 2). Among the four components, the financial ability is varying widely among the groups than other ability components. The financial

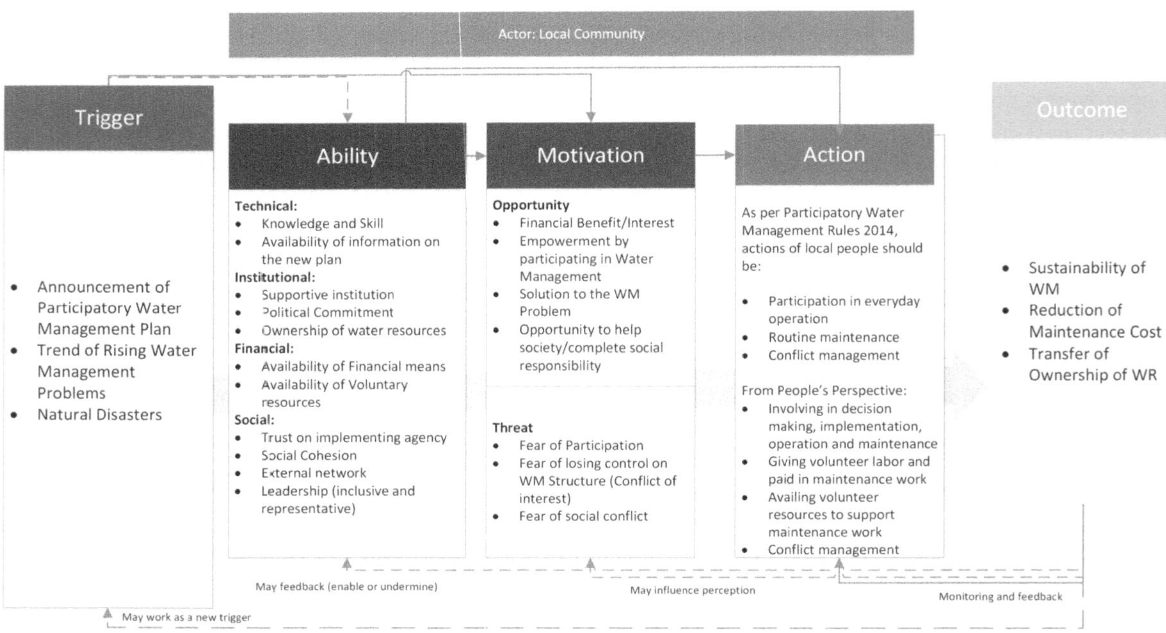

Fig. 1. | MOTA analytical framework for the societal adoptability of PWM.

Table 2 | Scores of MOTA-ability indicators and components.

Component	Indicator	MOTA score[a] for ability Scale 0 ~ 5 SF	MF	LF	AL	CSO-NGO	RF	SrmpF	Weights (0 ~ 5)[b]
Financial	Availability of financial means	0.5	1	1	1.3	2	1	0	3
	Availability of voluntary resources	2	3	2	3.8	3	3	4	4
Institutional	Supportive institution	1	1	2	1.8	4	1	3	5
	Ownership of water resources	1	4	3	1	3	3	3	5
	Political commitment	2	1	1	2	3	1	1	4
Social	Trust on implementing agency	1	2	1	2	3	1	0	4
	Social cohesion	2	4	3	3.5	4	4	2.5	4
	External network	1	2	3	3	5	1	3	3
	Leadership	1	1	2	1.8	3	2	4	4
Technical	Knowledge and skill	2	2	3	2.8	5	3	2	5
	Availability of information on the new plan	1	3	2	2	4	2	2	4

SF, small farmer; MF, medium farmer; LF, large farmer; CSO-NGO, civil society organization and non-governmental organizations; SrmpF, shrimp farmer.
[a]Score is a number quantifying the indicator, which was done by the experts following the MOTA operationalization method.
[b]Weights of each indicator were set by the experts. First, each group decided the weights independently, and they presented the weights to all, discussed, argued, and finally, all participating experts agreed to a common weight.

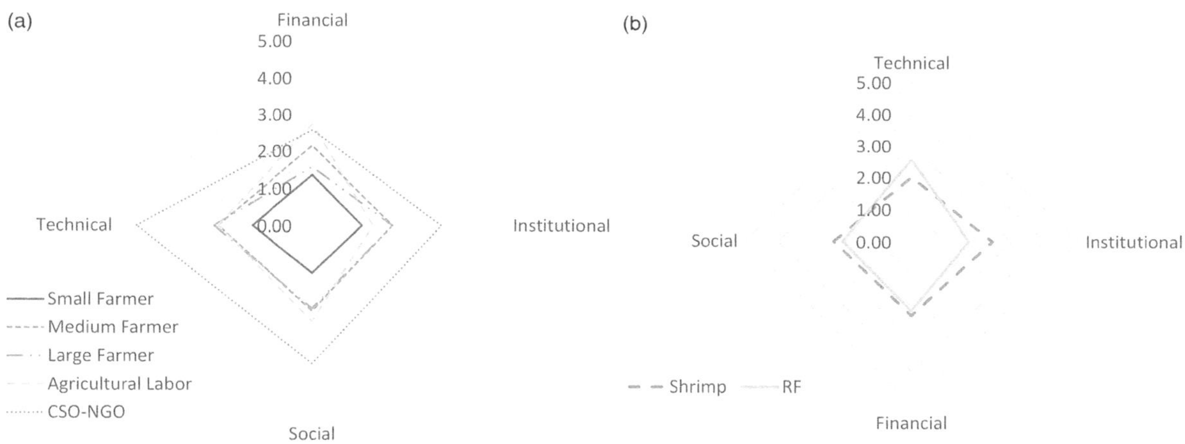

Fig. 2. | Status of four MOTA-ability components for different livelihood groups: (a) comparison among small farmer, medium farmer, large farmer, agricultural labor, and CSO-NGOs and (b) comparison between shrimp farmers and rice farmers (RF).

ability component is constructed by two indicators – availability of financial means and availability of voluntary resources. The availability of funds is very low in the case of all groups, and it varies between '0' and '2' among the groups (Table 2). Whereas the availability of voluntary resources is better, it varies between '2' and '4' among the groups (Table 2). The CSO-NGO shows a better score because, according to the experts, CSO-NGOs have a better capacity in mobilizing the community in voluntary work. Among the livelihood groups, agricultural labor shows surprisingly better financial capacity, because the availability of voluntary labor is higher in their community. According to experts, agricultural labors are mostly available in voluntary work than other livelihood groups.

Similar to financial capacity, the institutional capacities of livelihood groups are also very low, whereas the CSO-NGO shows better institutional capacity than others again. Among the livelihood groups, shrimp farmers show little better institutional capacity than others. While assigning the score, the expert groups are playing the role of shrimp farmers, arguing that in most of the cases, shrimp farmers are local elites who have strong political powers, affiliation, or association. Moreover, shrimp farmers have a very strong business association and network making them very connected with the government, which ultimately help them to influence the process of formulation of rules and regulations. Since their business is highly connected with water, they have higher ownership than others.

Social capacity is also very low in the case of all livelihood groups. In the case of CSO-NGO, social capacity is higher than livelihood groups due to their better external network, social cohesion, and leadership. The shrimp farmer also has higher leadership ability, but the values of other indicators are low.

Technical capacity is almost similar for most of the groups, CSO-NGOs and small farmer group are exceptions. CSO-NGO shows higher technical ability, and the small farmer group shows very low technical ability than others.

There are differences in abilities between the rice farmer group and the shrimp farmer group as well. The shrimp farmer outranked the rice farmers in the case of financial, institutional, and social ability, but the rice farmers surprisingly show higher technical ability. According to the experts, the rice farmers have better knowledge on local water resources, knowledge on operation and maintenance of small-scale water management structures which equip them better than shrimp farmers to participate in local water resources management.

Finally, the overall ability scores of livelihood groups, which are the average of four components of each group's ability and represented in percentage, are low varying between 40 and 46% with an exception of the small farmer group, which shows a very low ability score (27.33%) (Figure 3). The CSO-NGO has moderate to good ability, which is around 70% (Figure 3).

The overall ability scores subjectively reflect the chances of adopting the PWM plan by different stakeholder groups, which are unfortunately not very encouraging except the CSO-NGO group. Although this is neither the direct measurement nor the self-assessed value, the group of experienced experts participating in the validation and testing agreed to this result and argued that the result reflects the practical scenario of field. The test results also match with the subjective assessment of PWM case studies in Bangladesh carried out earlier by Buisson *et al.* (2017) and Dewan *et al.* (2014, 2015), where they found lack of financial capacity, weak institutional settings, and lack of quality in the participation implementation process of the implementing agencies, and a few social contexts (e.g. heterogeneity of the society, inequity, elite capture, and conflict of interest among the society) are causing inefficacy in PWM in the coastal community.

Apart from exploring and quantifying the adoptability, the MOTA indicators here are also relatable to the measure for enhancing the societal adoptability as well as the implementation feasibility of PWM. Taking the indicator values into account, the policymakers and implementing agencies can plan capacity development program targeting the low-performing indicators. Moreover, based on the indicator scores and overall MOTA scores of different stakeholder groups, capacity development programs can be customized for each stakeholder group. Thus, the extended MOTA framework informs policymakers and implementing agencies on how to close the implementation gaps and ensure the effectiveness of PWM while reforming under the BDP 2100 initiatives.

DISCUSSION AND REFLECTION

The test application of the MOTA indicator framework validates that the indicators are sensitive to the changes of stakeholder's capacity and thus confirms its readiness to be applied in field research with different stakeholder groups for assessment of the societal adoptability of PWM. The scoring technique and presentation used matches

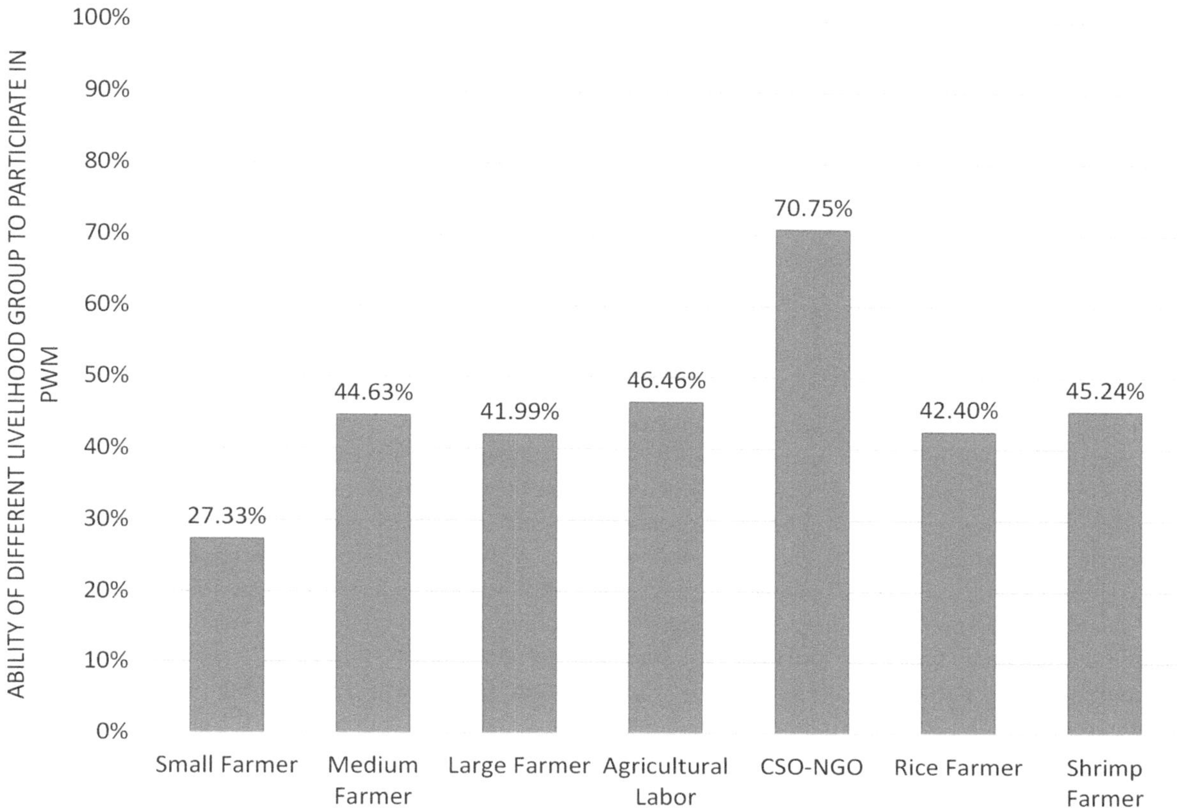

Fig. 3. | Ability of different livelihood groups to participate in local water management. *Note*: The overall ability score is the average of four ability components, which has been normalized and then represented in percentage.

with conventional ways of exploring the societal context, with relevant quantification techniques of qualitative assessment, and performance-oriented plan assessment tools. Therefore, this extended MOTA framework can be readily applied, along with other tools, for plan feasibility assessment.

In water management, many decision-makers and planners have an engineering background and feel comfortable with quantitative analyses to analyze problems and solutions. The quantification of the indicators in our presented MOTA framework aligns well with this thinking. In quantification, the MOTA framework relies on self-assessments through surveys or on expert judgements. Both have significant challenges and shortcomings, which cause some social scientists to be wary of quantified measures altogether. Our position is that these challenges in reliable quantification are real and that the limits of any (MOTA) study need to be clearly communicated in its presentation, but that such research is not impossible and can be useful – following earlier quantitative social studies cited above.

Our test application shows that the MOTA is quantifiable and it resonates with the mindset of these decision-makers and planners. By doing so, the MOTA framework offers a useful complement to other plan assessment tools and decision-making approaches that consider other planning dimensions, which are measurable and translated into numbers (as in cost–benefit analysis, multi-criteria analysis, and environmental impact assessment). Yet, at the same time, the MOTA framework also includes the thinking of social scientists, social workers, and

NGOs. That implementation is complex, political, and context-dependent and has different impacts on different societal actors. Moreover, by extending its scope to include assessment of measures related to institutional and social changes, it increases its application, enables communication, and broadens acceptance. Thereby, this research offers an additional planning tool to aid the implementation of the BDP 2100 in Bangladesh, with promises for similar applications elsewhere.

These first explorations further suggest that indeed the MOTA framework can be extended to also assess the feasibility of institutional and societal implementation measures. This is important, because, for the implementation of long-term goals and strategies, and a soft path or a soft implementation approach that considers necessary institutional reforms, social capacity enhancement, social acceptance, commitment, and adaptive learning is advocated (Brandes & Kriwoken, 2006; Korbee *et al.*, 2019). Our extended MOTA framework for exploring societal adoptability supports assessing such institutional and social implementation measures and the social adoptability of a plan. It supports planners in flagging the need for capacity development and institutional reforms to increase the implementation feasibility of the proposed measures.

Our extended MOTA indicator framework for PWM is contextualized for the case of policy development in Bangladesh. However, the indicators are based on a comprehensive literature study covering PWM experiences around the globe. On this basis, researchers in other countries might be able to start using (and adjusting) the framework for their scientific or planning interests. This does not mean that our MOTA framework can be used without testing in other contexts, testing to the local context is an essential part of using any analytical framework.

The remaining question is whether our extended MOTA indicator framework is also useful for assessing other institutional measures, like water pricing, decentralization of government institutions, regulatory changes for water pollution, and introduction of a new early warning system for a disaster. Most of the indicators identified in our extended MOTA framework for societal adoptability are specific to the issue of PWM. We expect that for other institutional measures, the indicators framework needs to be adjusted to that specific measure. Nevertheless, we think that the overall approach of this research (e.g. the approach of operationalizing MOTA, scoring, calculating, and testing) can be followed by others for assessing the adoptability of other institutional measures. Replicating this framework one-on-one to other contexts would do no justice to the complexity of implementing water policies and plans.

CONCLUSION

In this paper, we have presented our work on extending and testing the MOTA framework (Phi *et al.*, 2015; Korbee *et al.*, 2019; Nguyen *et al.*, 2019a, 2019b) for assessing the societal adoptability of an institutional measure, namely PWM in Bangladesh.

We conclude from our testing that the developed MOTA indicators indeed deliver insights into the societal capacity to participate, which informs policymakers and implementing agencies. Although our test results are limited to a relatively small group of experts, these expert estimates suggest that the financial, institutional, social, and technical abilities of societal actors are inadequate to support the successful and sustainable implementation of PWM. This implies that the capacity development of societal actors would be needed to make PWM successful. By detailing out the capacity (to participation) of different stakeholder groups into 11 indicator levels, the MOTA assessment framework opens up new information to design (and redesign) institutional measures that better fit the target stakeholders or to invest in the stakeholders to increase their adoptability of the proposed measures.

Our findings indicate that the MOTA framework is not issue bound and it can assess both large-scale infrastructural as well as institutional measures. However, this does not mean that the framework does not require any adjustment when to be used for different kinds of measures. Extending and adjusting the framework are possible but require care and rigor. We also conclude nevertheless that the MOTA framework is not bound in the Vietnamese planning tradition and can be used to assess and inform planning in other planning cultures. This does not mean that such planning cultures do not matter in how, when, and by who the MOTA framework is used in assessing plan implementation feasibility.

To sum up, our work shows: First, the (extended) MOTA framework can assess the implementation feasibility of strategic measures. Second, the MOTA framework can be used beyond its original geographical boundary. Third, it shows a rigorous approach to adjust it for other planning assessment purposes, showing how the MOTA framework can be extended by reviewing the literature, operationalizing indicators, and expert testing. And finally, our work shows that the results of these activities provide informed outcomes (e.g. insights of stakeholders' capacity to adopt an institutional change) that can support planners and decision-makers to improve the implementation feasibility of their plans. With these findings, we hope to contribute to the debate about, and further development of, analytical tools that help to include notions of implementation feasibility in plan and policy development, thus improving the quality of those plans and policies.

DATA AVAILABILITY STATEMENT

All relevant data are included in the paper or its Supplementary Information.

REFERENCES

Ajzen, I. (2002). Perceived behavioral control, self-efficacy, locus of control, and the theory of planned behavior. *Journal of Applied Social Psychology 32*(4), 665–683. https://doi.org/10.1111/j.1559-1816.2002.tb00236.x.

Arnstein, S. R. (1969). A ladder of citizen participation. *Journal of the American Planning Association 35*(4), 216–224. https://doi.org/10.1080/01944366908977225.

Ballester, A. & Mott Lacroix, K. (2016). Public participation in water planning in the Ebro River Basin (Spain) and Tucson Basin (U.S., Arizona): impact on water policy and adaptive capacity building. *Water 8*(7), 273. https://doi.org/10.3390/w8070273.

Bardhan, P. (2000). Irrigation and cooperation: an empirical analysis of 48 irrigation communities in South India. *Economic Development and Cultural Change 48*(4), 847–865. https://doi.org/10.1086/452480.

Brandes, O. M. & Kriwoken, L. (2006). Changing perspectives – changing paradigms: taking the 'soft path' to water sustainability in the okanagan basin. *Canadian Water Resources Journal 31*(2), 75–90. https://doi.org/10.4296/cwrj3102075.

Buisson, M.-C., Das, A. & Mukherji, A. (2017). Where river meets the sea. *Review of Market Integration 9*(1–2), 45–65. https://doi.org/10.1177/0974929217717528.

Carter, J. & Hill, G. (2007). Indigenous community-based fisheries in Australia. *Journal of Environmental Management 85*(4), 866–875. https://doi.org/10.1016/j.jenvman.2006.10.021.

Cavalcanti, C., Schläpfer, F. & Schmid, B. (2010). Public participation and willingness to cooperate in common-pool resource management: a field experiment with fishing communities in Brazil. *Ecological Economics 69*(3), 613–622. https://doi.org/10.1016/j.ecolecon.2009.09.009.

Desbureaux, S. (2018). *Creating the Commons? Social Cohesion and the Impact of Community Based Common Pool Resources Management in Madagascar*. Available at: https://sebastien-desbureaux.netlify.app/publication/voi/.

Dewan, C., Buisson, M.-C. & Mukherji, A. (2014). The imposition of participation? The case of participatory water management in coastal Bangladesh. *Water Alternatives 7*(2), 342–366.

Dewan, C., Mukherji, A. & Buisson, M.-C. (2015). Evolution of water management in coastal Bangladesh: from temporary earthen embankments to depoliticized community-managed polders. *Water International 40*(3), 401–4016. https://doi.org/10.1080/02508060.2015.1025196.

Dyer, J., Stringer, L. C., Dougill, A. J., Leventon, J., Nshimbi, M., Chama, F., Kafwifwi, A., Muledi, J. I., Kaumbu, J. M. K., Falcao, M., Muhorro, S., Munyemba, F., Kalaba, G. M. & Syampungani, S. (2014). Assessing participatory practices in community-based natural resource management: experiences in community engagement from Southern Africa. *Journal of Environmental Management 137*, 137–145. https://doi.org/10.1016/j.jenvman.2013.11.057.

Fogg, B. (2009). A behavior model for persuasive design. In *Proceedings of the 4th International Conference on Persuasive Technology – Persuasive '09, 1*, https://doi.org/10.1145/1541948.1541999

Franzén, F., Hammer, M. & Balfors, B. (2015). Institutional development for stakeholder participation in local water management-an analysis of two Swedish catchments. *Land Use Policy 43*, 217–227. https://doi.org/10.1016/j.landusepol.2014.11.013.

GED (2018). *Bangladesh Delta Plan 2100 (Bangladesh in 21st Century), Vollume 1: Strategy*. General Economic Division, Bangladesh Planning Commission, Dhaka, Bangladesh.

Heering, H. (2014). *Water Management Organisations – Comparative Analysis*. Blue Gold Program. Available at: http://bluegoldbd.org/wordpress/wp-content/uploads/2015/11/TR-09_-WMOs-Comperative-Analysis.pdf.

Hermans, L. M. & Thissen, W. A. H. (2009). Actor analysis methods and their use for public policy analysts. *European Journal of Operational Research 196*(2), 808–818. https://doi.org/10.1016/j.ejor.2008.03.040.

Hertin, J., Berkhout, F., Wagner, M. & Tyteca, D. (2004). *Are 'Soft' Policy Instruments Effective? The Link Between Environmental Management Systems and the Environmental Performance of Companies*. SPRU Working Paper Series. Available at: https://ideas.repec.org/p/sru/ssewps/124.html.

Hobbs, S. J. & White, P. C. L. (2012). Motivations and barriers in relation to community participation in biodiversity recording. *Journal for Nature Conservation 20*(6), 364–373. https://doi.org/10.1016/j.jnc.2012.08.002.

Jennewein, J. S. & Jones, K. W. (2016). Examining 'willingness to participate' in community-based water resource management in a transboundary conservation area in Central America. *Water Policy 18*(6), 1334–1352. https://doi.org/10.2166/wp.2016.005.

Jiménez, A., LeDeunff, H., Giné, R., Sjödin, J., Cronk, R., Murad, S., Takane, M. & Bartram, J. (2019). The enabling environment for participation in water and sanitation: a conceptual framework. *Water 11*(2), 308. https://doi.org/10.3390/w11020308.

Kapiriri, L., Norheim, O. F. & Heggenhougen, K. (2003). Public participation in health planning and priority setting at the district level in Uganda. *Health Policy and Planning 18*(2), 205–213. https://doi.org/10.1093/heapol/czg025.

Korbee, D., Nguyen, H. Q., Hermans, L. & Ho Long, P. (2019). Navigating the bureaucracy: an analysis of implementation feasibility for the Mekong Delta Plan, Vietnam. *Journal of Environmental Planning and Management 62*(9), 1545–1561. https://doi.org/10.1080/09640568.2019.1623014.

Krishna, A. (2002). *Social Capital, Community Driven Development, and Empowerment: A Short Note on Concepts and Operations*. Available at: http://documents1.worldbank.org/curated/en/511391468765604827/pdf/330770Kirshna01see0also0330741781.pdf.

Krueger, R., Tuler, S. & Webler, T. (2001). What is a good public participation process? Five perspectives from the public. *Environmental Management 27*(3), 435–450. https://doi.org/10.1007/s002670010160.

Kulsum, U. (2020). *Supporting Adaptive Delta Management: Systematic Exploration of Community Livelihood Adaptation as Uncertainty*. Delft University of Technology (TU Delft). https://doi.org/10.4233/uuid.

Kyamusugulwa, P. M. (2013). Participatory development and reconstruction: a literature review. *Third World Quarterly 34*(7), 1265–1278. https://doi.org/10.1080/01436597.2013.824653.

Lammerink, M. P., Bolt, E., de Jong, D. & Schouten, T. (1999). Strengthening community water management. *PLA Notes 35*, 21–28.

Macnaghten, P. & Jacobs, M. (1997). Public identification with sustainable development: investigating cultural barriers to participation. *Global Environmental Change 7*(1), 5–24. https://doi.org/10.1016/S0959-3780(96)00023-4.

Manca, A. R. (2014). Social cohesion. In *Encyclopedia of Quality of Life and Well-Being Research*. Springer Netherlands, pp. 6026–6028. https://doi.org/10.1007/978-94-007-0753-5_2739.

Mancilla García, M. & Bodin, Ö. (2019). Participatory water basin councils in Peru and Brazil: expert discourses as means and barriers to inclusion. *Global Environmental Change 55*, 139–148. https://doi.org/10.1016/j.gloenvcha.2019.02.005.

Marré, A. W. & Weber, B. A. (2010). Assessing community capacity and social capital in rural America: lessons from two rural observatories. *Community Development 41*(1), 92–107. https://doi.org/10.1080/15575331003661099.

Ménard, C., Jimenez, A. & Tropp, H. (2018). Addressing the policy-implementation gaps in water services: the key role of meso-institutions. *Water International 43*(1), 13–33. https://doi.org/10.1080/02508060.2017.1405696.

Nakagawa, Y. & Shaw, R. (2004). Social capital: a missing link to disaster recovery. *International Journal of Mass Emergencies and Disasters 22*(1), 5–34. Available at: https://goo.gl/ZsEHjI.

Nguyen, H. Q., Guerra, J. D. P., Korbee, D., Luan, P. D. M. H., Dung, T. D., Hung, D. Q., Loc, H. H. & Hermans, L. (2019a). *Manual for Application of the MOTA Framework: Theory and Practice*. Available at: https://strategic-delta-planning.un-ihe.org/sites/strategic-delta-planning.un-ihe.org/files/quan_mota_manual_-_version_1.pdf.

Nguyen, H. Q., Korbee, D., Ho, H. L., Weger, J., Thi Thanh Hoa, P., Thi Thanh Duyen, N., Dang Manh Hong Luan, P., Luu, T. T., Ho Phuong Thao, D., Thi Thu Trang, N., Hermans, L., Evers, J., Wyatt, A., Chau Nguyen, X. Q. & Long Phi, H. (2019b). Farmer adoptability for livelihood transformations in the Mekong Delta: a case in Ben Tre province. *Journal of Environmental Planning and Management 62*(9), 1603–1618. https://doi.org/10.1080/09640568.2019.1568768.

OECD (2008). *Handbook on Constructing Composite Indicators: Methodology and User Guide*. Organization for Economic Co-operation and Development (OECD), Paris. Available at: https://www.oecd.org/std/42495745.pdf.

Özerol, G. & Newig, J. (2008). Evaluating the success of public participation in water resources management: five key constituents. *Water Policy 10*(6), 639–655. https://doi.org/10.2166/wp.2008.001.

Perkins, P. E. E. (2011). Public participation in watershed management: international practices for inclusiveness. *Physics and Chemistry of the Earth 36*(5–6), 204–212. https://doi.org/10.1016/j.pce.2010.02.004.

Phi, H. L., Hermans, L. M., Douven, W. J. A. M., Van Halsema, G. E. & Khan, M. F. (2015). A framework to assess plan implementation maturity with an application to flood management in Vietnam. *Water International 40*(7), 984–1003. https://doi.org/10.1080/02508060.2015.1101528.

Pineda, V., Chowdhury, A., Marandy, J., Hossain, S., Bentvelsen, K., Bijlmakers, H. & Smits, D. (2013). *Outcome of Functionality Assessment for Water Management Organizations Polders 22, 30, 43/2D and 43/2F (Issue 1)*. Blue Gold Program. Available at: http://bluegoldbd.org/wordpress/wp-content/uploads/2015/11/TR-03_WMO-Functionality-Assessment-Polders-2230-43.2D-and-43.2F.pdf.

Pineda, V., Chowdhury, A., Marandy, J., Hossain, S., Smits, D., Heering, H., Bentvelsen, K., Bijlmakers, H. & Smits, D. (2014). *Outcome of Functionality Assessment for Water Management Organizations Polders 22, 30, 43/2D and 43/2F (Issue 3)*. Blue Gold Program. Available at: http://bluegoldbd.org/wordpress/wp-content/uploads/2015/11/TR-03_WMO-Functionality-Assessment-Polders-2230-43.2D-and-43.2F.pdf.

Pritchett, L., Woolcock, M. & Andrews, M. (2013). Looking like a state: techniques of persistent failure in state capability for implementation. *Journal of Development Studies 49*(1), 1–18. https://doi.org/10.1080/00220388.2012.709614.

Razzaque, J. (2009). Public participation in water governance. In *The Evolution of the Law and Politics of Water*. Springer Netherlands, pp. 353–371. https://doi.org/10.1007/978-1-4020-9867-3_21.

Rowe, G. & Frewer, L. J. (2000). Public participation methods: a framework for evaluation. In *Science Technology and Human Values*. Sage Publications, Thousand Oaks, CA, Vol. 25, Issue 1, pp. 3–29. https://doi.org/10.1177/016224390002500101.

Samaddar, S., Yokomatsu, M., Dayour, F., Oteng-Ababio, M., Dzivenu, T., Adams, M. & Ishikawa, H. (2015). Evaluating effective public participation in disaster management and climate change adaptation: insights from northern Ghana through a user-based approach. *Risk, Hazards & Crisis in Public Policy 6*(1), 117–143. https://doi.org/10.1002/rhc3.12075.

Samaddar, S., Okada, N., Choi, J. & Tatano, H. (2017). What constitutes successful participatory disaster risk management? Insights from post-earthquake reconstruction work in rural Gujarat, India. *Natural Hazards 85*(1), 111–138. https://doi.org/10.1007/s11069-016-2564-x.

Sovacool, B. K. (2011). Hard and soft paths for climate change adaptation. *Climate Policy 11*(4), 1177–1183. https://doi.org/10.1080/14693062.2011.579315.

Stanghellini, P. S. L. (2010). Stakeholder involvement in water management: the role of the stakeholder analysis within participatory processes. *Water Policy 12*(5), 675–694. https://doi.org/10.2166/wp.2010.004.

Sultana, F. (2009). Community and participation in water resources management: gendering and naturing development debates from Bangladesh. *Transactions of the Institute of British Geographers 34*(3), 346–363. https://doi.org/10.1111/j.1475-5661.2009.00345.x.

Swapan, M. S. H. (2016). Who participates and who doesn't? Adapting community participation model for developing countries. *Cities 53*, 70–77. https://doi.org/10.1016/j.cities.2016.01.013.

Tosun, C. (2000). Limits to community participation in the tourism development process in developing countries. *Tourism Management 21*(6), 613–633. https://doi.org/10.1016/S0261-5177(00)00009-1.

van Maasakkers, M., Duijn, M. & Kastens, B. (2014). Participatory approaches and the role of facilitative leadership. In: *Risk-Informed Management of European River Basins. The Handbook of Environmental Chemistry*. Brils, J., Brack, W., Müller-Grabherr, D., Négrel, P. & Vermaat, J. E. (eds). Springer, Berlin, Heidelberg, Vol. 29, pp. 295–319. https://doi.org/10.1007/978-3-642-38598-8_11.

van Staveren, M. F., Warner, J. F. & Shah Alam Khan, M. (2017). Bringing in the tides. From closing down to opening up delta polders via Tidal River Management in the southwest delta of Bangladesh. *Water Policy 19*(1), 147–164. https://doi.org/10.2166/wp.2016.029.

Von Korff, Y., Daniell, K. A., Moellenkamp, S., Bots, P. & Bijlsma, R. M. (2012). Implementing participatory water management: recent advances in theory, practice, and evaluation. *Ecology and Society 17*(1). https://doi.org/10.5751/ES-04733-170130.

Webler, T. & Tuler, S. (2001). Public participation in watershed management planning: views on process from people in the field. In *Human Ecology Review*. Society for Human Ecologym, Vol. 8, pp. 29–39. https://doi.org/10.2307/24707126.

Wesselink, A. (2016). Trends in flood risk management in deltas around the world: are we going 'soft'? *International Journal of Water Governance 4*(4), 25–46. https://doi.org/10.7564/15-ijwg90.

Wesselink, A., Paavola, J., Fritsch, O. & Renn, O. (2011). Rationales for public participation in environmental policy and governance: practitioners' perspectives. *Environment and Planning A: Economy and Space 43*(11), 2688–2704. https://doi.org/10.1068/a44161.

First received 1 December 2020; accepted in revised form 13 June 2021. Available online 9 July 2021

© 2022 The Authors From Knowledge and Capacity Development to an Implementation Science: doi: 10.2166/wp.2021.248

Water specialist as andragogist: the application of learning theory in capacity development for improved water management

Bobby Russell * and Bouke Ottow

Deltares, Boussinesqweg 1, 2629 HV Delft, The Netherlands
*Corresponding author. E-mail: bobby.russell@deltares.nl

BR, 0000-0003-0941-6718

ABSTRACT

This paper argues for greater application of adult learning theory in capacity development for water agencies in developing countries. Although the need for capacity development is clear, we believe its delivery, especially through 'technical assistance' projects, is of limited effectiveness. In the authors' experience, this limited effectiveness is due to the mismatch between intention and delivery. This may be due to inappropriate objective setting, lack of focus in training delivery and poor cultural awareness, reinforced by an unchallenged, ineffective model of teaching. A more collaborative and adult learner focused approach can help to define the learner's current situation and identify the learning methods that suit them best. Although the effectiveness of capacity development has many factors, the diligent application of well-tested adult learning practices and culturally appropriate learning methods can give positive results. This paper describes five principles for improving capacity development in developing countries. Those principles are: structural alignment of learning, culturally appropriate learning methods, active learning methods, learning trajectories and 'water specialist as andragogist'. The paper describes and evaluates three case studies of capacity building activities, where these principles were applied either in a deliberate structured approach or through happy accident.

Key words: Adult learning, Andragogy, Capacity development, Cross-cultural teaching, Curriculum, Learning styles, Training, Water management

HIGHLIGHTS

- This paper argues that water specialists become 'andragogists' and apply adult learning theory for institutional capacity strengthening.
- This paper demonstrates the application of five principles which improve institutional capacity strengthening: structural alignment of learning, culturally appropriate learning methods, active learning methods, learning trajectories and 'water specialist as andragogist'.

INTRODUCTION

In this paper, we state that technical experts in the field of water, like in a knowledge institute such as Deltares, can improve the effectiveness of their capacity development activities with the application of good learning theory and practices. Although the need for institutional capacity building in international development is clear, its effective delivery is a well-reported challenge (World Bank, 2008; Vallejo & Wehn, 2016). These challenges range from human and budgetary resources, enabling environment, legislative frameworks, to transparency and good governance (Alaerts & Kaspersma, 2009). We believe, based on our collective experience of capacity building projects in developing countries, that the limited effectiveness of the uptake of skills and knowledge (i.e. the training element of capacity development) may be the result of the mismatch between intention and delivery, often due to inappropriate objective setting, lack of focus in training delivery, poor cultural awareness and a reliance on poorly defined buzzwords (such as 'learning-by-doing' or 'knowledge transfer') over effective practices and more regard for the content than for the actual learning needs. From over 20 years of collective experience in the sector, we have observed that typical donor funded capacity development projects often set ambitious learning objectives with limited resources. The result is training that skips vital precursor competencies and is simply delivered too fast, considering the skill level of the organisation's staff. In other cases, objectives are set in a considered way, but there is a disconnect between the objective, training method and what the learners actually do in their organisation. Within typical capacity development projects, the people delivering training are often technical specialists with no prior knowledge of adult learning principles, yet are expected to deliver training. Without this prior awareness, these technical experts are less able to reflect on what the learners' needs, their cultural and education backgrounds, and thus naturally gravitate to a subconscious model of teaching they are most familiar with. Often that is not the model that is most appropriate in the training courses that are needed. The result we observe is a one-way passive transmission of 'know-how' or 'knowledge transfer' that fails to recognise how adults learn. It is a model of the 'assembly line worker' in which the 'expert does – the beneficiary receives' thus a traditional passive didactic teaching approach (Plueddemann, 1986). A more collaborative and learner focused approach can help to define the learner's current situation and identify the learning methods that suit them best. This also relates to modern human resources approaches where staff development (learning) is enabled through 'coaches' and 'mentors' rather than directive 'managers' (Lombardo & Eichinger, 1996).

Although adult learning theory is well known and established in education institutions, we argue that for capacity development in international water projects, it presents an important innovation. Such arguments for considered adult-orientated learning approaches have also been made before, for example Vallejo & Wehn (2016) referring to the extensive work of Tough (1971) on adult education in the work place, while this paper discusses their application. It is also recognised that 'capacity development' requires more than just 'training', although the terms are often used as synonyms, or are at least overlapping. Quoting the UNDP definitions (Wignaraja, 2009), we understand 'training' to be the 'Systematic educative process by which one learns new skills' and 'capacity development' the 'continuous planning carried out to advance a person's career based on experience and on any training undertaken to upgrade qualifications or to acquire new ones'. We also understand that capacity development can refer to the development of capacities of organisations. In our understanding, capacity development always has at least some focus on individuals, and capacity development always contains 'training' whether formal and planned or a more informal 'learning-by-doing'. This paper focuses on the training element of capacity development whether for individuals or part of a longer-term career and organisational learning trajectory.

There are many water management practitioners who do apply learning theories, but, overall, from our experience, we believe that too many do not. Moreover, we believe that there is a general unawareness in the water management profession of the practical use that can be made of those theories. Additionally, the number of academic studies addressing the usefulness of learning theories in capacity development is quite small, and studies on its effectiveness are even less. The reason for this might be the difficulty of rigorously testing a pre-conceived theoretical framework in the reality of training courses defined for projects. The intention of this paper is not to rigorously test theoretical frameworks, but to describe candid observations from actual implementation, presented as three case studies. Thus raising more general awareness among water professionals of the application of learning theories and showing where rigorous testing by researchers might be applied to strengthen the wider knowledge base. The Moldova–Kenya–Rwanda case reflects the internal learning trajectory of one of the authors as a water specialist working on international development projects aimed at capacity building. The Training of Trainers (TOT) in Bangladesh case was part of the Training and Transfer of Tools Work Package of the Joint Cooperation Programme Bangladesh – The Netherlands, in which the authors participated as employees of Deltares. The Deltares 'Pizza course' cases, on the design and delivery of technical training courses, were organised by the authors to train technical colleagues on andragogical principles and practice.

METHODOLOGY

This section describes five principles for improving the capacity of government water agencies and knowledge organisations in developing countries. In several of these cases, the principles are founded on well-established learning theory applied in universities and human resources management, but are less well known in capacity development in the water sector. Moreover, quantitative research into their effectiveness has been performed in regard to their application in tertiary education and adult learning but not in the context of capacity development projects. Those principles are:

1. 'the water specialist as andragogist',
2. constructive alignment of learning,
3. active learning methods,
4. learning trajectories and
5. culturally appropriate learning methods.

The paper describes and evaluates three case studies of capacity building activities, where these principles were applied either in a deliberate structural approach or through happy accident.

The case studies are:

Case study 1: The Moldova–Kenya–Rwanda learning trajectory
Case study 2: TOT in Bangladesh
Case study 3: 'Pizza' courses in Deltares, the Netherlands

'Water specialist as andragogist'

Adult education theory (andragogy) is not an area most water specialists are familiar with, but the subject provides an insight into why training approaches commonly fail. The authors observed that technical consultants on capacity development projects, who have a responsibility to achieve learning objectives, complain that 'simple concepts' are not grasped as if that is the fault of the learner. But technical consultants need to reflect on how adults learn and on the years of formal-training and self-learning that resulted in their own current skills and knowledge before trying to impart that knowledge. This will make it easier to identify key precursor knowledge and skills and thus build a realistic and step-wise learning programme. Knowledge of adult learning theory also empowers technical consultants to address the learners' motivations and needs in a more focused way. Six key assumptions, drawn from Knowles' theory on adult learning (Knowles, 1980), are:

- *Need to know:* To be motivated to learn, adults need to know the reason for learning something. Thus, in capacity building, the 'why' must be clearly described to the learner.
- *Foundation on prior experience:* Adults reflect on their earlier experience while learning. Thus, in capacity building, we must solicit this experience and build on it but never ignore it.
- *Readiness:* Adults are most interested in learning techniques and tools that have immediate relevance to their daily work. Therefore, the 'why' must be linked to the learner's immediate day-to-day needs in the work place.

- *Self-concept:* Adults feel the need to be responsible for decisions regarding their education, therefore when designing the learning objectives and curricula the learner should be involved in the process.
- *Problem based orientation:* Adult learning is often more successful if it presents a solution to problems the learners recognise rather than being content-orientated.
- *Motivation:* Adults respond better to internal as opposed to external motivators. The former is intrinsic to curious, self-directed learners who prefer to explore a subject than be told simply how to do something.

Knowles' principles provide an insight into the motivations of adult learners, especially how it relates to human resources development in the workplace, and thus a better approach to designing capacity development programmes.

Learning trajectories

A learning trajectory (Figure 1) plots the path an organisation takes in acquiring skills and knowledge and in many ways mirrors the curricula structuring of higher education course programmes with the identification of overarching learning objectives and the building blocks of learning activities necessary to attain the objectives. It is presented in the diagram as a smooth steady idealised curve, but the reality of adult learning in the workplace means it is often a more stop–start progress with some backwards steps. It is also very much dependent on the

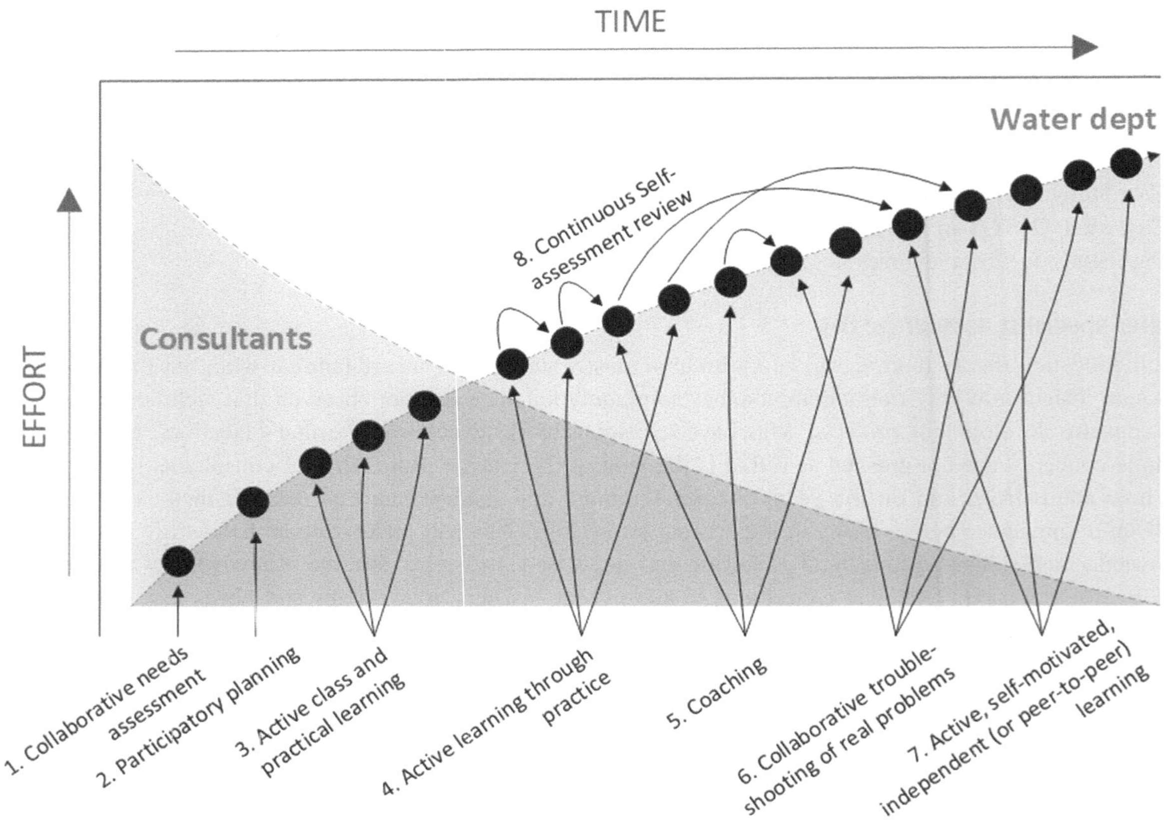

Fig. 1. | Conceptual learning trajectory in capacity building (Source: Russell, 2018).

pace of each learner in an organisation, so each individual staff member and the organisation itself would have its own learning trajectory.

Tying into Knowles principles, the foundation of a learning trajectory in capacity development is the learners need to know, problem orientation and internal motivation. The learning process should be clearly and convincingly laid out in a coherent 'story' that learners can visualise and evaluate for themselves. If they see a real value in the trajectory, where their prior knowledge, day-to-day tasks and learning styles are taken into account then there is a better chance of success. One of the key aspects is active learning through practice. The implementation of it is easily overlooked but requires significant effort to create mid- to long-term opportunities for learners to practice. Built into those opportunities should be informal assessment by the learner where, with a coach, they are able to reflect on their own learning, identify what they need to do and how best to achieve it. The overall aim of the approach is that the learners become more independent and self-driven learners as they move through the process.

Constructive alignment of learning

Constructive alignment is a practice from academic teaching proposed by Biggs (2003) which links constructivist theory of experiential learning and outcome-based evaluation of formal education. The practice (Figure 2) makes a clear link between:

- the teacher's intent for a training course (that is the 'learning objective'),
- the learners' activity (that is the 'doing' or experiential part of the course) and
- the learners' assessment (that is the achievement of the learning objective).

The key is that the stated verb in the learning objective (that is what are the learners expected to be able to DO) should be reflected in the learning activity and the assessment.

Simply put, if the trainer expects the learners to be able to DO something after the training, then that training must include the learner DOING that activity. Talking about it or demonstrating it is not sufficient. Although in capacity development of water agencies there is no examination in the traditional sense, we do need to assess whether the training has achieved the stated objectives. In its most basic form the assessment is: are the learners using the skills learned?

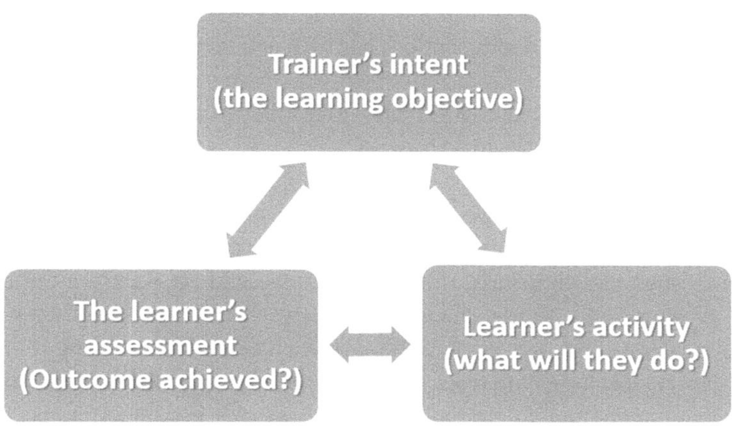

Fig. 2. | Constructive alignment (Source: adapted from Biggs (2003)).

Table 1 | Commonly assumed knowledge retention of different learning activities.

	Learning activity carried out	% retention after 1 month
Active	Teaching others	80
	Practical exercise	75
	Group discussion	50
Passive	Demonstration	30
	Audio-visual	20
	Reading	10
	Lecture	5

Active learning methods

Active learning is a planned series of actions or events designed to provide the opportunity for participants to process, apply, interact and share experiences as part of an education process. It has been defined as follows: 'learning takes place through the active behaviour of the student: it is what she/he does that she/he learns, not what the teacher does'. After Tyler (1949), Table 1 shows the often quoted and discussed figures associated with Dales' pyramid of learning (Letrud, 2012) that are commonly used to indicate the knowledge retention following a learning activity.

Such retention figures are well known in academia, and although they are challenged, our anecdotal experience confirms the ineffectiveness of 'traditional' lecturing and reading-based learning, so called passive methods. Therefore, organisational capacity building should apply well-aligned active learning activities. As a means of creating a 'teaching others' learning environment, the TOT approach and the sharing of existing local knowledge are very effective. The theory behind this is that if the learners teach the material, and thus have to explain it to others, they (have to) internalise it more than when 'just' understanding would be sufficient.

Culturally appropriate learning methods

People learn differently, as is illustrated, for instance, by the different Kolb learning styles (Kolb, 1984). Another learning style differentiation is that of field-dependent and field-independent learners, as developed by Herman Witkin in 1954 (Witkin *et al.*, 1977). Later, the term 'field-dependent' was replaced by the term 'field-sensitive'. Where field-independent learners are good at abstract analytical thought and tend to be more individualistic, field-sensitive learners have a more global perception and are more sensitive and attuned to social environment, with highly developed social skills. Field-independent learners are more intrinsically motivated, whereas field-sensitive learners are more responsive to social reinforcement (Bennett, 1990).

Research suggests that learning style is related to world view that certain learning styles tend to be predominant in certain cultures (Bennett, 1990). Following Hofstede's cultural dynamics (Hofstede *et al.*, 2010), one can expect that in cultures that score high on 'power distance' and low on 'individualism', many learners would prefer a 'field-sensitive' learning style, whereas in cultures with low scores on 'power distance' and especially with high scores on 'individualism', one can expect that many learners would prefer a field-independent learning style. Although there are, of course, individual differences, one would expect more 'field-sensitive' learners, for instance, in Asia than in Europe or North America, see Figure 3. In this figure, we see, for example, a low score on individualism for China (20) and Japan (46) and a high score for Germany (67) and USA (91).

For those technical experts that originate from cultures with dominant field-independent learning styles that train people from cultures with dominant field-sensitive learning styles, it is helpful to be aware of some field-sensitive behaviours which would help adjust training methods (after Bennett, 1990):

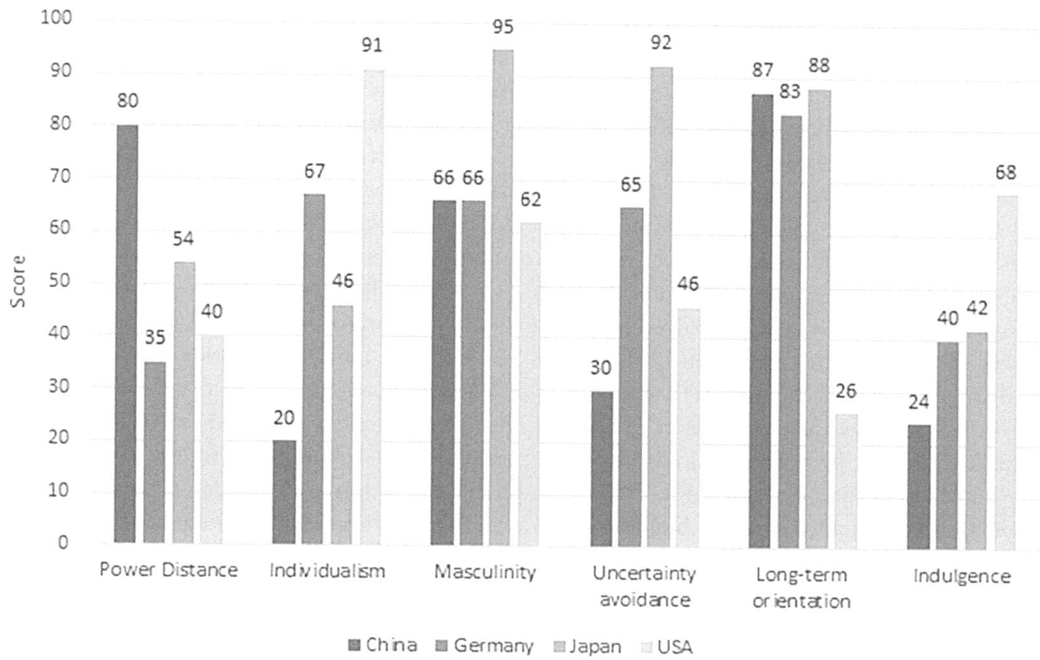

Fig. 3. | Scores on Hofstede's cultural dimensions for China, Germany, Japan and USA (Fan *et al.*, 2017).

Field-sensitive learners:

- Like to work with others to achieve a common goal
- Like to assist others
- Are sensitive to feelings and opinions of others
- See the trainer's taste and personal experiences as leading
- Seek guidance and demonstration from a trainer
- Seek rewards which strengthen relationship with a trainer

A field-sensitive training:

- Makes purpose and main principles obvious, rules explicit
- Personalises the curriculum (related to interests and experience of learners)
- Humanises the curriculum: use of narration, humour and drama
- Encourages learning through modelling
- Encourages the cooperation and development of group feelings
- Includes informal class discussions

The application of the principles

As a demonstration, we describe three case studies where the principles were applied either by design or by happy accident. The case studies are the reflections of the authors as the consultants implementing the capacity building activities.

Case study 1: The Moldova–Kenya–Rwanda learning trajectory

The following case reflects the internal learning trajectory of one of the authors as a water specialist working on international development projects aimed at capacity building. The trajectory tracks the progress of the individual from a pure technical specialist to a 'hydrologist as andragogist' and how that was applied.

The author worked on a capacity building project in the Republic of Moldova for a period of 3 years (2011–2015), delivering a series of training courses for government institutions in technical matters concerning hydrology and water resources planning. They were technical topics and approached in a technical manner, structuring the activities as classroom activities followed by field or computer practicals, much as the author had been instructed at university. The experience made the author question why some learners progressed better than others. Subsequently, the author was involved in a capacity building project in Kenya for a technical university in the delivery of water-related education (2012–2016). The capacity building included the delivery of a condensed workshop, delivered by the IHE Institute for Water Education, on curricula development and pedagogical skills. Subjects that most water specialists are not exposed to. The author realised that many of the principles had, in fact, been applied in Moldova, but not consciously, and furthermore, it raised an interest in pedagogy and then eventually andragogy. Subsequently, this led to a more considered, structured and adult learning-orientated approach to capacity building in Rwanda for a government water agency (2015–2018).

The delivery of training in the Moldova project was distinctly focused on the technical topic of the project and was not learner-orientated. That is to say, the learning objectives did not utilise Bloom's taxonomy (Bloom *et al.*, 1956) in their formulation. Although this is a minor point it meant that inherently the 'topic' was the focus, whereas the focus should be on what the learners are to 'do'. This misalignment was recognised in the Kenya project and subsequently addressed when designing capacity building curricula. The formulation of the learning objectives had to be more thoroughly discussed with the beneficiaries, to make sure the objectives matched the expectations, as well as truly represented the actual needs and more difficultly, matched the available resources. We realised that constructive alignment (Biggs, 2003) was needed, but the evaluation of the achievement of objectives remained a challenge as traditional university style exams are inappropriate and, although self-assessment has a value, it is not always objective. We found that the insertion of simple text questions (both formally and informally) into the learning trajectory to test understanding and skills attained is a useful way for the instructors to evaluate the achievement of learning objectives and adjust the trajectory accordingly. The testing of it was done by adapting Mazur's ConcepTest approach (Crouch & Mazur, 2001), originally developed and rigorously tested for evaluating the attainment of core understanding in university physics education. In the context of the capacity development in Kenya and Rwanda, they were applied as oral questions in the field or presented as classroom questions (see Figure 4). These 'test questions' focus on the core understanding needed to achieve the originally stated learning objective and also initiate self-assessment by the learner.

The trajectory through the three projects demonstrated that 'active learning' is not as simple as first thought. It's often treated as 'learning-by-doing', and interpreted as the simple process of 'watch me, now you do it', although technically 'active', the cognitive processes may be missing. Initially in Moldova, the 'watch me, now you do it' approach was used for most practical exercises. Almost immediately, it became apparent that some individuals could not critically evaluate what they were doing, nor self-identify where their mistakes were. The latter is a crucial skill in the workplace. At a certain point the learners began to discuss the problem among themselves, one individual started to go through the process they had just learned, describing it visually for their colleague, who quickly started to grasp the concept. The author initially assumed that it was simply a language barrier problem, but years later, courtesy of the Kenya project, realised that this 'happy accident' had been extensively documented in academic teaching practices as 'peer-to-peer' learning (Crouch & Mazur, 2001).

Example ConcepTest 1:

Consider the flow in a river as it enters a constriction. What will happen to the water surface elevation between point A and B?

A – It increases.

B – It decreases.

C – It remains the same.

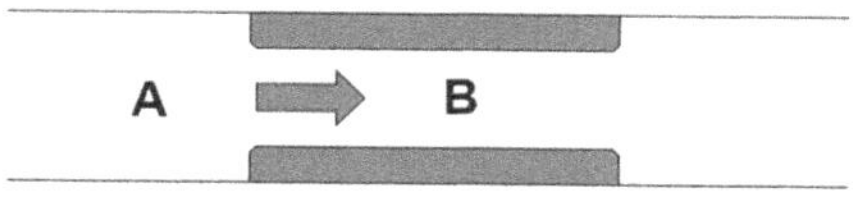

Fig. 4. | Application of the ConcepTest™ for water management.

The idealised learning trajectory presented above (Russell, 2018) was a conscious effort applied in all the three projects to look at the learners as a group and as individuals, linking the learning objectives to a programme of adult-orientated active learning activities. The entirety of the curricula package was discussed with the water department to solicit feedback but also to empower them in directing their own learning. This approach led to some interesting findings. Firstly, senior government officials wished for a focus on sophisticated water modelling software, whereas the actual learners from the government departments were enthusiastic participants in a foundation course on the application MS Excel, a relatively simple but endlessly applicable software. They recognised a real value for their day-to-day jobs. This identification and delivery of training for foundation skills proved very effective and was borne out by feedback from participants.

The initial classroom and practical instruction included greater focus on skills and knowledge that the learners needed for the day-to-day jobs, and thus they themselves recognised as relevant. Opportunities were created to allow them to practise these skills, especially for field work and practical work, although much of the follow-up practice was with gradually reducing supervision. The technical consultants would return periodically to review progress and help with trouble shooting. Group discussion was encouraged, and there was a strong focus on explicitly asking the learners to reflect on their own learning, identifying how it fit with their prior knowledge and concepts and they were asked to describe what problems they encountered and how they addressed it. This was also used as opportunities to introduce ConcepTests™ for self-assessment.

The experience from Moldova, Kenya and Rwanda demonstrated that there are strong differences between the instructor (from western Europe) and the learners (from eastern Europe and East Africa). Culture is not simply an issue of religion, ethnic or language but almost a deep unconscious learning style, created by years of experience. Moldova was very much rooted in Soviet education systems, while Rwanda and Kenya produce graduates rooted in a didactic teaching system that traditionally does not create an environment where open critical evaluation is encouraged. Instructively in all three countries, learners at different times asked 'so is this the 'correct' answer?' without evaluating what might constitute a 'correct' answer in the realm of water management. The author found, for example in Kenya, that senior university lecturers and vocational educationalists found concepts such as 'learners teaching others' alien to their current practices. Therefore, it was crucial to introduce new learning methods iteratively and in small steps. A successful approach in Kenya was during the introduction of new pedagogical methods. The learners were given a strict assignment as follows:

'In small groups (<5 people), select one of the following scientific papers for critical evaluation. You have 1 hour to read and review the paper, formulate a presentation of no more than 3 slides and present the paper in 3 minutes. The presentation shall include the title, brief description of the research hypothesis and research question and the results and your evaluation.' (Russell & Paron, 2015). After completing the assignment, the participants had to (1) reverse engineer our learning objectives for their assignment and (2) identify what active learning methods they employed. The results produced a large amount of productive debate on what were culturally alien approaches such as learning-by-teaching and active learning. In the space of 2 h of contact time, the learners had a good understanding of these concepts and how to apply them in practice. Whereas when these concepts were previously presented as simple bullets in a presentation they had struggled to grasp them.

Case study 2: TOT in Bangladesh

This TOT, in 2019, was part of the Training and Transfer of Tools Work Package of the Joint Cooperation Programme Bangladesh – The Netherlands (JCP). Participants were trained in the core concepts of capacity development within the context of Integrated Water Resources Management (IWRM). The 18 participants of the TOT were all employees of the Bangladeshi research institutes (CEGIS and IWM) responsible for the training of different actors in the joint research projects. The objectives of the training were to (Ottow *et al.*, 2019):

- Provide participants with the basic tools and knowledge to design, conduct, monitor and evaluate effective training within the context of IWRM and
- Support the participants in the development of a training plan for their own project.

Based on the course objectives, the learning objectives were specified for five specific aspects: Principles of IWRM and its application to Bangladesh; Training principles; Teaching methods; Training development; Presentation. Below are the specific learning objectives for the training principles and the training development aspects.

Training principles:

- Participants understand that for effective learning, we must focus on the active behaviour of the students, not that of the teacher.
- Participants understand the concept of learning style and its importance in relation to training design; they know their own preferred learning style and the possible consequences for their selection of teaching activities.
- Participants can identify the main actors in and around their projects that need to be trained.
- Participants can identify these actors' project-related training needs.
- Participants can translate these training needs into training objectives, formulated in actor behaviour.

Training development:

- Based on the identified training needs, the available time and resources and the general context of their project(s), participants can select/identify/formulate realistic training objectives.
- Participants can identify appropriate evaluation methods for the selected objectives.
- Participants can identify appropriate teaching methods related to these objectives and evaluation methods.
- Participants know the elements of effective presentations.
- Participants can make a realistic, coherent training plan.

During the training course, the participants worked on the development of training plans for the six projects within the JCP during the different learning activities (among which a serious game) and on the last day gave mini-lessons and presented draft training plans for their respective projects. Even though the assignments and presentations were in groups (varying from three to six members), and we could not evaluate for every participant, the participants clearly had added some andragogist skills. After an introduction on the first day, the participants identified their own experiences through a YES/NO psychological survey (Kolb test) and marked themselves to find their own learning approaches/mental models (metaphors) and learning styles. This identified the suitable metaphors/learning styles for the intended training within their projects. During the activities, the participants then related the different learning methods to the different learning styles. In their presentations of the training plans, three of the six groups also included the Kolb learning styles as an important ingredient for determining the appropriate learning methods.

The self-assessment scores of the TOT showed that the participants' knowledge of teaching had 'much increased'. The aspect of learning style was mentioned several times in the self-assessment and four participants mentioned the learning styles/cycle of Kolb as having been most helpful in achieving the objectives of the TOT.

The participants showed an understanding of the importance of constructive alignment of learning. In developing the training plans for their respective projects, the participants first identified the main actors/ stakeholders in their project and their training needs, then translated these into learning objectives, then the method of evaluation and then, finally the learning activities in line with the objectives and the evaluation method. In the self-assessment, the participants stated that their skills in drafting a training plan had 'much increased'. Some indicated that although they drafted training plans before, they now have a more systematic way to do it.

The learning trajectory is the one aspect that was paid least attention to. The six groups identified the different target groups and their training needs quite well but translated all these into learning objectives and activities for one training event per target group, mostly consisting of 3 days and not (yet) in learning trajectories. After an introductory lecture on teaching/training methods, the participants identified in groups the suitability of different teaching/learning activities for learners with different learning styles. In their plans, they intentionally considered and included active learning activities.

Besides presenting the training plans on the final day, the groups also gave a short demonstration lesson on the subject matter. All six groups did that in a lively, original and activating way: discussion in small groups; assignment in small groups; test/game and a demonstration in a movie.

In the self-assessment, the participants indicate that their knowledge of teaching had 'much increased', while teaching skills, and especially the awareness of the need for it, had on average also 'much increased'. Some participants indicated that they needed more practice, indicating the value of a 'learning trajectory' over one-off training.

In the design and delivery of the TOT itself, we tried to include many different types of learning activities that would cater to different learning styles and certainly also to field-sensitive learners. There were interactive lectures, group assignments, presentations, tests, a serious game and the assignments to design a lesson plan and design and deliver an interactive learning activity.

Even if in 4 days the participants did not learn all the necessary insights and skills about learning and teaching for effective training courses, we do believe that this relatively short training of trainers gave a big boost to their understanding of the necessity to apply good learning theory and gave them ideas and already a little practice on how to apply those.

Case study 3: 'Pizza' courses in Deltares, the Netherlands

Deltares, as a knowledge institute for water and sub-surface, has always implemented internal training courses. Sessions are typically from 4 pm till 8 pm and when these courses started years ago, pizza was ordered and delivered for the participants. The intention of the courses is to train other colleagues in new water management software and techniques, thus ensuring that the organisation is abreast of the latest developments in the subject. Participation is mainly in staff member's own time and thus completely based on their intrinsic motivation and daily job needs, thus linking back to Knowles' principles of adult learning (Knowles, 1980).

Deltares is also required to share this knowledge and skills with external water agencies both in the Netherlands and internationally. So often, these internal courses form the basis of capacity strengthening activities for other organisations. Given the demand for such courses, technical specialists expressed interest in improving training methods, and thus the authors have regularly advised on the design and delivery of training courses organised by technical colleagues for international projects. Based on the remarks we received from those technical colleagues (i.e. 'more people should know this'), we initiated pizza courses on 'effective design of training courses in international projects'. That is effectively 'training the trainers' but also addressing an organisational need for more water specialists acting as andragogists.

Since 2014, we organised four pizza courses on effective training design, with an average number of 15 participants. Most of these pizza courses consisted of two sessions of 4 h each. In the first sessions, most of the educational principles were presented, tests (Kolb learning style) and short small group exercises done. In between the two sessions, the participants designed their own training programme, either individually or in duos. These were presented and discussed in the second session, in which also the cross-cultural aspects were introduced.

In the pizza courses, the main emphasis was on good learning theory and the application of it in the training courses. The participants were challenged to design an effective curriculum for their anticipated training courses and thus to think clearly about training needs (what their students should be able to do after the training, building on their existing knowledge and skills), learning styles, learning objectives and appropriately aligned learning activities. This included theory and tests on learning styles (such as Kolb below) to become aware of one's own preferred learning style and of the existence of different ones and the implications of it for learning/teaching activities.

Figure 5 shows the outcome of a Kolb learning style self-assessment. At the top of each 'kite', the score for 'activist' learning style is, at the right: 'thinker', at the bottom: 'theorist' and at the left: 'pragmatist'. The index refers to the number of questions answered in one of the four categories. These four 'kites' show the different preferred learning styles of these four people: the top-left 'kite' shows a preference for 'pragmatist', the top right one for 'activist', the bottom right one for 'thinker' and the bottom left one for 'theorist'. The application of this kind of test at the very preliminary stage of a capacity building project would help inform a more appropriate training programme.

The main focus in the discussion of the participants' training plans was on the consistency between overall purpose, learning objectives, learning activities and evaluation. Although we did mention the importance of active learning methods, especially for certain types of learning styles, we did not spend time to practise or discuss in depth. In most cases, our colleagues had a single training event as case, not a whole learning trajectory. By discussing the different target groups, their characteristics and their training need; however, we did touch on the necessity to look at the trajectory as a whole.

Besides the learning styles theory and test, we also presented theories on cultural differences, with a cultural values test (see Figure 6) to make participants more aware of the potential cultural differences in their trainees. Results shown are for a person characterised as more 'event-orientated' than 'time-orientated' (Figure 6, top left), more 'person-orientated' than 'task-orientated' (Figure 6, top right), more 'holistic thinking' than 'dichotomistic thinking' (Figure 6, bottom left) and more 'achievement-focused' than 'status-focused' (Figure 6, bottom right). As a group we discussed the implications that this could have for capacity development. What does it mean for your capacity development activity, for instance, when your learners are more event-orientated than time-orientated? Or more person-orientated than task-orientated? One can expect that they will less respect deadlines for a given assignment and that the relationship between learners and trainer becomes very important. Also, what does it mean for your learning activity if you as the trainer have the opposite values?

Although the degree in which the insight and skills were obtained differs per participant, and we have not systematically evaluated the transfer after the pizza courses, we could see from the results of the assignments that the approximately 60 colleagues had applied improved andragogical skills.

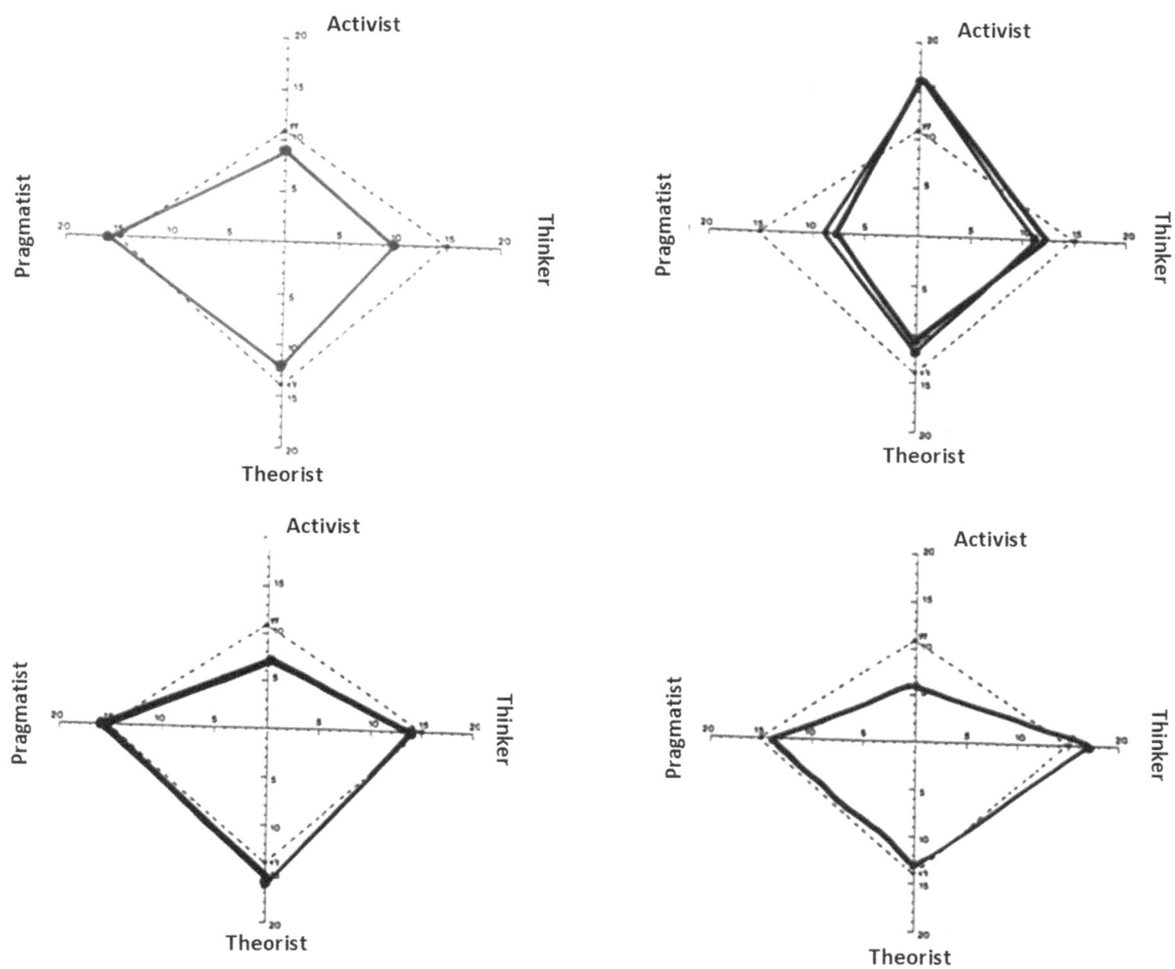

Fig. 5. | Some Kolb learning style tests outcomes ('kites') of participants in Deltares pizza courses 2014 and 2015.

DISCUSSION

In our own practice and in the experiences we obtained in training technical experts as trainers, we found that it is possible in a relatively short time to help them realise the importance of andragogical principles and to apply those to their technical training activities (Ottow *et al.*, 2019). We also found that once the attention and time is there, they are tremendously creative in developing appropriate, active, learning activities. Yet there remain profound challenges that are not necessarily remedied by the application of adult learning practices. Nor are they possible for capacity building consultants to directly resolve. These include the often very short timeframes for capacity development under international development projects. Additionally, water institutions in developing countries still have limited resources and are thus losing well-qualified (even newly trained) staff to the private sector.

There are challenges that can be influenced, and therefore we encourage the international donors, who formulate capacity development projects to recognise the importance of adult learning theory and (cultural) differences

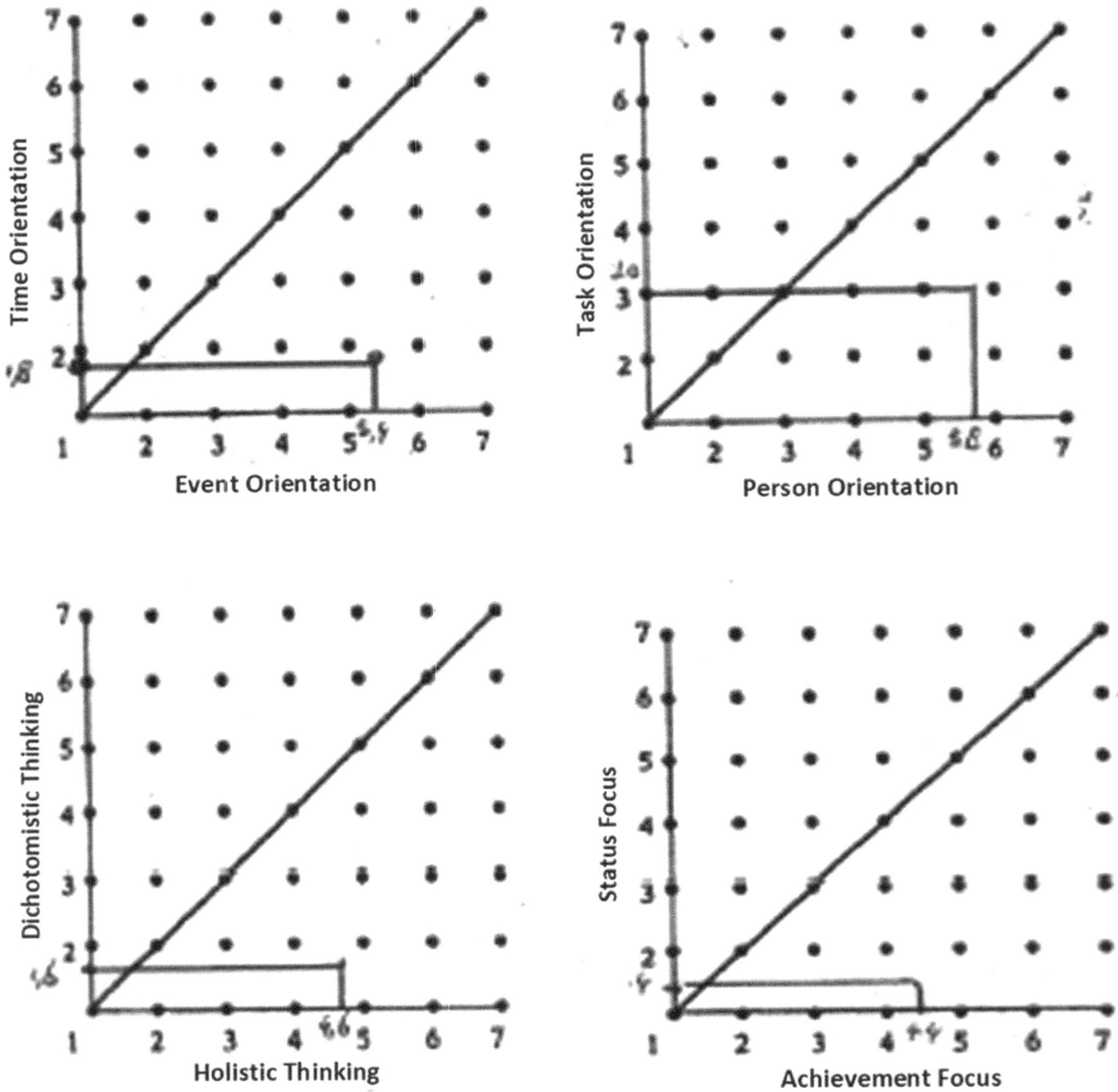

Fig. 6. | A result of the Lingenfelter cultural values test: time versus event orientation; task versus person orientation; dichotomistic thinking versus holistic thinking; status versus achievement focus (after Lingenfelter & Mayers, 1986).

in learning styles when formulating project terms of reference, and that the scoring of proposals rewards better approaches, such as those outlined here.

A further challenge is the rigorous assessment of the approaches described in this paper. The authors acknowledge that the anecdotal evidence presented here needs to be substantiated through further research. Such research, the authors argue, should focus on actual capacity building projects for water agencies, with a carefully considered monitoring and evaluation framework. Although the monitoring and evaluation of capacity development remains a well-documented challenge in itself (World Bank, 2008; Watson, 2010; Vallejo & Wehn, 2016), there is a wealth of knowledge on approaches to appropriately assess the achievement of capacity development objectives. Furthermore, the authors recognise that they were not impartial observers of the process but active participants; therefore, rigorous assessment needs a more independent perspective.

The authors recognise that the approaches they applied are not necessarily the most complete suite of learning approaches. There are other adult learning approaches such as micro-learning. Leong *et al.* (2020) presents micro-learning as a growing approach with the following key benefits: better retention of concepts, better engagement for learners, improving learners' motivation, engaging in collaborative learning and improving learning ability and performance. Micro-learning is also presented as a suitable means for continuous professional development within organisations via e-learning platforms (Buchem & Hamelmann, 2010). Given both the restrictions imposed by the COVID-19 pandemic, capacity development is changing rapidly and the combined micro-learning/e-learning approach may become a widely applied adult learning approach for the modern time/attention poor work environment. This has the added advantage of the learning material being available to the organisation long after a capacity development project has ended. This may present the new frontier in the adult learning/capacity development paradigm.

CONCLUSIONS

In conclusion, the principles applied in the three case studies point to a way forward in capacity development, connecting the worlds of learning theory and consultant water specialists. Although there remain many challenges, the authors pose the following:

1. That those responsible for the preparation of capacity development projects for water management institutes in developing countries create terms of reference with realistic objectives.
2. That capacity development projects for water management institutes monitor achievement of learning objectives through constructively aligned indicators.
3. That the same experimental research rigour is applied to capacity development and achievement of learning objectives, as it is for university education and adult learning methods.
4. That institutions that are active in developing technical capacity stimulate their technical staff to develop their andragogical skills.

DATA AVAILABILITY STATEMENT

All relevant data are included in the paper or its Supplementary Information.

REFERENCES

Alaerts, G. & Kaspersma, J. (2009). Progress and challenges in knowledge and capacity development. In: M. W. Blokland, G. J. Alaerts, J. M. Kaspersma & M. Hare, eds. *Capacity Development for Improved Water Management*. CRC Press, London.

Bennett, C. I. (1990). *Comprehensive Multicultural Education: Theory and Practice*, 2nd edn. Allyn and Bacon, Boston.

Biggs, J. B. (2003). Aligning Teaching and assessing to course objectives. In: *Teaching and Learning in Higher Education: New Trends and Innovations*. University of Aveiro, Aveiro, Portugal, 13–17 April 2003. Available from: https://www.researchgate.net/publication/241251310_Aligning_teaching_and_assessing_to_course_objectives

Bloom, B. S., Engelhart, M. D., Furst, E. J., Hill, W. H. & Krathwohl, D. R. (1956). *Taxonomy of Educational Objectives: The Classification of Educational Goals. Handbook I: Cognitive Domain*. David McKay Company, New York.

Buchem, I. & Hamelmann, H. (2010). Microlearning: A Strategy for Ongoing Professional Development, ELearning Papers, N° 21. P.A.U. Education, S.L., Barcelona.

Crouch, C. H. & Mazur, E. (2001). Peer instruction: ten years of experience and results. *American Journal of Physics 69*, 970–977.

Fan, Z., Huang, S. & Alexander, W. R. J. (2017). Do national cultural traits affect comparative advantage in cultural goods? *Sustainability 9*(7), 1–16.

Hofstede, G. H., Hofstede, G. J. & Minkov, M. (2010). *Cultures and Organizations: Software of the Mind: Intercultural Cooperation and its Importance for Survival*. McGraw-Hill, New York.

Knowles, M. S. (1980). *The Modern Practice of Adult Education: From Pedagogy to Andragogy*. Cambridge Adult Education Company, New York.

Kolb, D. A. (1984). *Experiential Learning: Experience as the Source of Learning and Development*. Prentice-Hall, NJ.

Leong, K., Sung, A., Au, D. & Blanchard, C. (2020). A review of the trend of microlearning. *Journal of Work-Applied Management 13*(1), 88–102.

Letrud, R. (2012). A rebuttal of NTL Institute's Learning Pyramid. *Education 133*(1), 117–124.

Lingenfelter, S. G. & Mayers, M. K. (1986). *Ministering Cross-Culturally – An Incarnational Model for Personal Relationships*. Baker Book House, Grand Rapids.

Lombardo, M. M. & Eichinger, R. W. (1996). *The Career Architect Development Planner*. Lominger, Minneapolis.

Ottow, B., Monowar Duti, B., van der Vat, M. & Sardjoe, N. (2019). *Report Training of Trainers I*. Joint Cooperation Programme (JCP), Bangladesh – the Netherlands.

Plueddemann, J. E. (1986). Metaphors in Christian education. *Christian Education Journal 7*(1), 39–47.

Russell, B. (2018). *An Adaptive Approach to Capacity Building for Integrated Water Resources Management*. (Unpublished).

Russell, B. & Paron, P. (2015). *Pedagogical Skills Training Materials for Technical University of Kenya*. (Unpublished).

Tough, A. (1971). *The Adult's Learning Projects: A Fresh Approach To Theory and Practice In Adult Learning*. Ontario Institute for Studies in Education, Ontario.

Tyler, R. W. (1949). *Basic Principles of Curriculum and Instruction*. University of Chicago Press, Chicago.

Vallejo, B. & Wehn, U. (2016). Capacity development evaluation: the challenge of the results agenda and measuring return on investment in the global south. *World Development 79*, 1–13.

Watson, D. (2010). Measuring capacity development. combining the 'best of two worlds' in monitoring and evaluation of capacity development. In: J. Ubels, N.-A. Acquaye-Baddoo & A. Fowler, eds. *'Capacity in Practice'*. Earthscan, London and Washington DC, pp. 251–263.

Wignaraja, K. (2009). *Capacity Development: A UNDP Primer*. UNDP, New York, USA.

Witkin, H. A., Moore, C. A., Goodenough, D. R. & Cox, P. W. (1977). Field-dependent and field-independent cognitive styles and their educational implications. *Review of Educational Research Winter 47*(1), 1–64.

World Bank (2008). *Using Training to Build Capacity for Development – An Evaluation of the World Bank's Project-Based and WBI Training*. Independent Evaluation Group, The World Bank, Washington, DC.

First received 9 November 2020; accepted in revised form 22 June 2021. Available online 23 August 2021

© 2022 The Authors From Knowledge and Capacity Development to an Implementation Science: doi: 10.2166/wp.2022.235

Assessing design principles for climate services training courses: educational design principles assessment of six C3S Blended Training courses within the Copernicus Climate Change Service

Maria del Pozo [a], Judith Gulikers [b], Erik van Slobbe [a,*], Perry den Brok [b] and Fulco Ludwig [a]

[a] Wageningen Environmental Research, Wageningen University and Research, Wageningen, Netherlands
[b] Social Sciences Group, Wageningen University and Research, Wageningen, Netherlands
*Corresponding author. E-mail: maria.delpozogarcia@wur.nl

Md, 0000-0002-4708-0760; JG, 0000-0002-9318-312X; Ev, 0000-0003-0499-2281; Pd, 0000-0002-4945-763X; FL, 0000-0001-6479-9657

ABSTRACT

The current climate service creation practice misses clear provider–user collaborations, and this presents a challenge for the educational design of capacity-building programs. This study analyses the formation of educational principles in six training courses aimed at tailored climate services. The design principles are analyzed using the constructive alignment and three curriculum perspectives as analytical frameworks. Three main issues were identified: overambitious one-size-fits-all learning goals; the role of a case study in overcoming the lack of knowledge and skills; and ambiguity in assessments. These issues guided the implementation for improvements in the courses and need to be addressed in creation processes for user-tailored climate services in general by the wide community of climate service providers and users. Our findings reflect the tendency to insufficiently involve users in the creation of climate services and in capacity building more specifically. Although we use examples in the water sector and link them to collaborative processes in water governance, our findings potentially have implications for other sectors where collaboration between users and providers is needed as well. It also highlights not only the usefulness of educational and pedagogical disciplines as a pillar of capacity building but also their active inclusion in the design and implementation of climate services.

Key words: Capacity building, Climate services, Educational design, Tailored climate services

HIGHLIGHTS

- Understanding the nature of tailored climate services is the first step into introducing its characteristics into design principles for capacity building.
- Including collaborative and transdisciplinary processes in capacity building programs potentially supports development of tailored climate services.
- Producers and users are very heterogeneous, and there is a need to further understand them in order to develop a tailored capacity building.

INTRODUCTION

Climate change has an important impact on the water sector. Climate extremes will become more frequent and intense in the future, resulting in more and intense floods and droughts. In order to facilitate adaptation, climate services are developed to provide tailored information. These climate services are supposed to provide scientific climate information to enhance users' knowledge to support their decisions and actions (Vaughan *et al.*, 2018). Hecht (1980) recognized the value of climate service's value even 40 years ago; in those days these services were driven primarily by national meteorological agencies and climate research institutes as climate information providers (Vaughan & Dessai, 2014). But, researchers observe that climate services' influence on the user's decision and actions are limited, and they refer to a 'usability gap' (Lemos *et al.*, 2012). Lourenço *et al.* (2016) attribute this usability gap to climate service providers' science-driven or supply-driven inclination. This usability gap results in a growing need for tailored climate services with user-demand-driven approaches that consider socio-economic aspects and adaptation options (Dilling & Lemos, 2011; Kirchhoff *et al.*, 2013; Lourenço *et al.*, 2016). The gap calls for user involvement and a shift towards a collaborative knowledge creation process known as co-production (Vincent *et al.*, 2018).

The use of collaborative approaches in climate service design is growing. Advances are also being made in unraveling stakeholders' relationships and influences to identify common challenges and solutions (García *et al.*, 2019; Rojas *et al.*, 2020). Co-production as a collaborative process helps to develop actionable knowledge that assists decision-making (Vincent *et al.*, 2018). Co-production not only requires an iterative and interactive collaboration but learning within and between users and providers (Nicolescu, 1997; Mauser *et al.*, 2013; Schuck-Zöller *et al.*, 2018). The results of co-produced climate services are described as 'demonstrably usable' knowledge developed through interactive and iterative scientific processes (Lemos *et al.*, 2012; Kuusaana & Bukari, 2015). Karpouzoglou *et al.* (2015) distinguish between first- and second-generation climate services. The difference is that second-generation climate services are based on co-production (Karpouzoglou *et al.*, 2015), making scientific information usable for decision-making and creating user-driven products. Vincent *et al.* (2018) add to this description that co-production of climate services leads to a result that is decision-driven, process-based and time-managed, developed within an inclusive, collaborative, and flexible process. Although co-production of climate services is discussed in different terms, it is commonly agreed that co-production is a learning process to create user-driven climate services and, therefore, tailored to the user needs. Yet, most literature on co-production in climate service is fragmented and context-specific, making a complete understanding of the nature of tailored climate services difficult (Weichselgartner & Arheimer, 2019).

In general, climate services distinguish between users and producers. While researchers agree that different users have different needs, the literature commonly does not differentiate between types of users (Swart *et al.*, 2017). In practice, producers and users are heterogeneous. Similarly, to specific fields like water governance, climate services require multiple layers and scales of stakeholders: from international, national, regional, and/or local authorities, regulators, and civil society from public and private sectors, including a wide range of scientific disciplines (Weichselgartner & Arheimer, 2019; Rojas *et al.*, 2020). Weichselgartner & Arheimer (2019) distinguish climate service providers and producers in providing climate service to the user. Climate service providers produce raw climate data, while the producer processes or adds value to the raw data to provide the required information to the user. The producer that adds value is also called a purveyor (Máñez *et al.*, 2014). The World Meteorological Organization (WMO) uses different terminology. It considers the National Meteorological and Hydrological Services (NMHSs) as climate service producers and distinguishes between different types of users: 'users', 'intermediary users', and 'end-users'. While the 'end-users' are the decision-makers to whom the information is provided to, 'users' interpret, analyze and process climate and weather information, adding their sector-specific knowledge to produce a useable, tailored, and integrated climate service. The 'intermediary users' are the National Meteorological and Hydrological Services' partners, working together to produce climate services (WMO, 2010). Therefore, the distinction and understanding in the literature of users and providers/producers are ambiguous terms, embracing a vast background and expertise of different levels and sectors.

The WMO (2010) states that capacity building is a critical climate service pillar for improving information provided to users. Capacity building for climate services includes, in its curriculum, climate data management, monitoring and prediction, service delivery, and product communication to users. Although WMO differentiates between users, intermediate users, and end-users, its capacity-building pillar suggests to distinguish between climate service providers and capacity building for users (Domingos Freires, 2016). Capacity building for users focuses on understanding climate variability and change, and interpretation and use of climate services provided. Capacity building for climate service providers aims to deliver accurate and reliable information and communication to the user (WMO, 2012). This division potentially maintains the breach between users and providers rather than integrated within the co-production of the information, making it questionable if these capacity-building programs can foster more tailored climate services.

The heterogeneity within climate service producers and users, combined with the unbalanced emphasis on the supply of data and limited understanding of the nature of tailored climate services, does present a challenge for capacity building and its design. Therefore, the main research question is 'What are the design principles for capacity building for tailored climate services?' We use the term climate services as the tailoring of decision and climate-relevant information that provides actionable knowledge to relevant users (Weichselgartner & Arheimer, 2019). We also understand capacity building as obtaining and improving skills and knowledge through various means such as workshops, trainings, dialogues, and others. But answering the question is not only relevant for capacity building as such, but the outcomes have wider implications because their findings suggest the principles for changing the cognitive basis and skills required for the creation of climate services in general.

The empirical basis for this study is a capacity-building project in the frame of the Copernicus Climate Change Service (C3S) program. The C3S project aimed at improving the quality and usability of tailored climate services by offering courses in EU member countries. The courses proved to be an opportunity to test, integrate, and adjust principles for capacity building for tailored climate services, given its international, open access, and wide-scale around Europe. This paper presents an evaluation of the educational design principles from the six first-year C3S Blended Training courses. The results from the assessment of those six first courses in the first year allow us to define concrete implications for the principles and correspondingly adjust and test them again on the following ones. These implications may improve capacity building to promote the provision of user-relevant climate information and, therefore, tailored climate services in the considered most climate-sensitive sectors – agriculture, health, disaster reduction, and water (World Meteorological Organization, 2010).

To understand the working of design principles, we use the constructive alignment approach (Biggs, 2003) and the three curriculum perspectives of Van den Akker (2004) as analytical frameworks. The consistencies and differences within the course design as well as its comparison with the current challenges on capacity building for climate services offer a unique opportunity to draw conclusions for future capacity building and its inclusion on the design of tailored climate services, together with ideas for future research.

Before describing the study, the next sections present the context of the study, which is the C3S Blended Training courses and the two conceptual frameworks used to analyze the educational design of these courses.

Empirical basis: the C3S Blended Training courses

The 'C3S Blended Training courses' are part of the C3S. The European Centre for Medium-Range Weather Forecasts (ECMWF) operates the C3S on behalf of the European Union. The C3S is a climate service that offers quality-assured information about the past, current, and future states of the climate in Europe and worldwide. The C3S provides monthly climate bulletins with maps and guidance on current climate change indicators. The C3S also provides diverse demonstrator projects to present key sectorial themes to address climate-related issues that businesses or communities are facing across Europe. The main component of the C3S is the Climate

Data Store (CDS). The CDS is a one-stop platform that provides free available climate data to enable climate change adaptation and mitigation strategies for policy-makers and businesses (ECMWF, no date a).

The User Learning Services (ULS) is C3S' capacity building component. The ECMWF outsourced the ULS development to Wageningen University and Research (WUR). WUR won the service by a tender (ECMWF, 2017) and has developed the ULS as a service contract from 2018 until 2021. The service contract included two other project partners: Leeds University, which is tasked with developing an initial training requirement report and with monitoring and evaluation, and the Royal Netherlands Meteorological Institute (KNMI), which is tasked with developing online lessons and assisting the courses as master trainers. The ULS offers online lessons and blended training courses through a free access learning platform. Individuals can choose between 28 online lessons, which are subdivided into two categories: understanding of climate data (e.g., the data discovery, climatic data sources, and sectorial applications uncertainty) and use of the CDS (ECMWF, no date b). These lessons are included as learning activities within the 'C3S Blended Training courses'. Additionally, to the individual learning, 30 training courses have been organized on location in different European countries.

The 'C3S Blended Training courses' are designed with a blended approach that combines online learning lessons, two webinars, and a 1-day place-based classroom on location with hands-on activities. An essential educational activity in the courses was developing a case study aimed at designing a climate service for an adaptation challenge related to the learner. The course encouraged learners to bring their own adaptation challenges as a basis for this case study. The case studies encompassed a variety of fields relevant to their context. However, most of the case studies not only focused on climate-sensitive sectors, and specifically on the nexus between water and agriculture, but also related water to other sectors like energy and ecology. Case study's titles ranged from 'Seasonal weather forecasts for agriculture' in Italy, 'Characteristics of the wave regimes in the Baltic Sea' in the training in Latvia, 'Effects of climate change on river fish populations of Gipuzkoa' in the training in Spain to 'adaption to climate change in the Tagus River' in Portugal. The final case studies developed by learners are presented at the course end and submitted through the learning platform.

The 'C3S Blended Training courses' were designed by WUR curriculum designers and implemented by master trainers from the service contract and local trainers from the in-country courses. Those courses were internally monitored and evaluated by the University of Leeds and annually evaluated by trainers. While this evaluation from trainers was based on the informal exchange of insights, this paper aims to develop an educational, scientific-based analysis of the first-year courses.

Conceptual framework

This study assessed the six courses performed in the first year of the service contract using two conceptual frameworks relevant for educational design: the constructive alignment model of Biggs (1996) and the curriculum perspectives of Van Den Akker (2013). Constructive alignment is a pedagogical principle that defines and aligns what students should be learning, how they are learning, and how they are assessed (see Figure 1). If a learning outcome aims to develop a specific skill, activities need to align with practising this skill and assessment methods should allow learners to demonstrate they have met the *intended* result. Constructive alignment thus allows for identifying the consistencies and inconsistencies among the learning goals, activities, and assessments.

A learning goal is a clear statement of what a learner is expected to do or know after following the educational entity (e.g., module, training, etc.). Teaching is designed to engage students in learning activities that enhance their abilities for attaining those learning outcomes. Assessment tasks are designed to measure how well learning goals have been achieved. Assessments may include traditional exams and also presentations, assignments, or group work (Biggs, 1996).

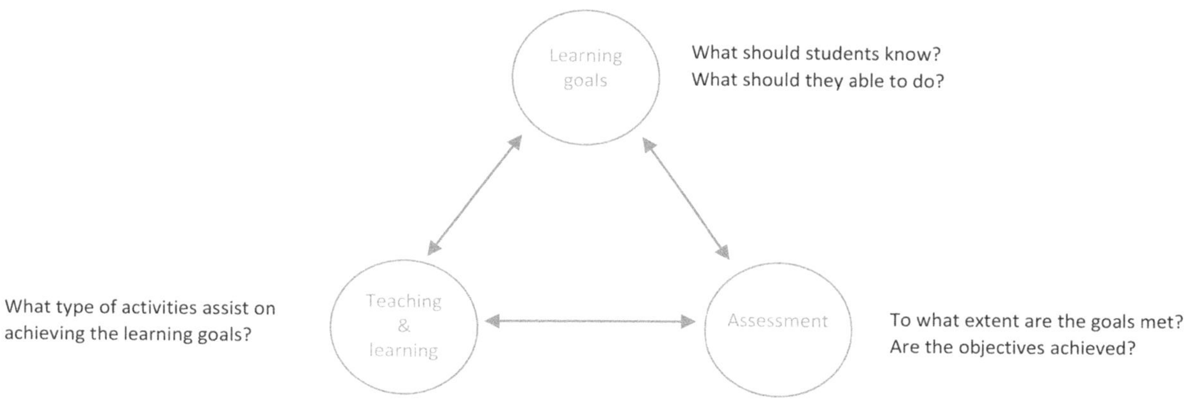

Fig. 1 | Adaptation constructive alignment adapted from Biggs (2003).

The second framework used in the analysis of the C3S Blended Training courses contains the curriculum perspectives of Van den Akker. Van Den Akker (2013) understand 'curriculum' in the context of education as the course, trajectory, or 'learning plan'. Given this definition, a curriculum is represented in different forms: the *intended*, *implemented*, and *attained* curriculum. Conventionally, the *intended* form relates to how the curriculum developers designed (*intended*) the curriculum to be (in various roles). The *implemented* form refers to the world of teachers and trainers, that is, how teachers implemented the training, and the *attained* relates to learners and students and how they experienced their learning in the course. To understand how to change a curriculum, Van Den Akker (2013) highlight the need to clarify those forms to define inconsistencies between them and change the curriculum.

METHODS

The six C3S Blended Training courses are the first courses developed within the ULS between September 2018 and July 2019 around Europe: The Netherlands, Italy, Spain, Serbia, Latvia, and Portugal. These first courses had the same intended curriculum design in terms of goals, activities, and assessments, though their implementations could differ to a certain degree. However, for the *implemented and attained curriculum*, we did not aim to analyse individual training courses but instead to identify commonalities in the implemented designs. In this study, we distinguish between local trainers and master trainers. Master trainers are experts in the field of climate change. Local trainers come from the origin country from the training, and their involvement in the learning activities depends on their background and expertise. Two master trainers involved in the service agreement contract implemented each course together with one to three local trainers from each in-country course. The courses had an average of 27 participants. Each participant submitted an individual case study proposal at the beginning of the course, of which four to six cases per course were selected to develop as a team. The case studies in The Netherlands were not considered in this study as they were not submitted to the learning platform, and there was no record of them. A summary of the courses is shown in Table 1.

Different data sources and methodologies were used to assess the *intended*, *implemented*, and *attained* goals, activities, and assessments (see Table 2). The *intended* domain is based on the written formal intentions reflected in documents and materials from the curriculum developers (Van Den Akker 2013). The *intended* goals, activities, and assessments were evaluated by comparing contract service documents, including the call for tenders from the ECMWF and the service contract proposal from WUR. The results from the document analysis were

Table 1 | Summary of the C3S Blended Training courses implemented during 2018–2019

In-country course	City for the face-to-face event	Period	Local trainers	Learners	Individual proposals case study	Final case studies submitted
The Netherlands	Wageningen	September 2018	7	31	–	–
Serbia	Belgrade	October 2018	2	27	17	17
Italy	Bologna	November 2018	3	27	22	7
Spain	San Sebastian	March 2019	2	34	14	6
Latvia	Riga	April 2019	2	12	3	3
Portugal	Lisbon	June 2019	2	30	17	5

Table 2 | The relation between the curriculum perspectives, methods, and data collected

Curriculum perspective	Representation	Description	Methods	Data collected
Intended	Ideal	Vision	Interview's analysis	Four working package leaders (curriculum designers) One ECMWF project representative
	Formal/ written	Specified intentions in documents and material	Document analysis	Tender description ECMWF Service contract proposal WUR
Implemented	Operational	Actual teaching and learning process	Document analysis	Report training requirements Leeds University Five training programs Five trainers reports
	Perceived	Curriculum interpreted by trainers	Interview's analysis Evaluation reports written by trainers	Nine local trainers Four master trainers
Attained	Experiential	Learning experiences perceived by trainers	Interview's analysis	
	Learned	Resulting outcomes of learners	Document analysis	45 case studies 44 individual proposals

discussed in individual interviews with the four proposal developers from WUR as they designed the curriculum. The *implemented* goals, activities, and assessments were extracted from reported training requirements by Leeds University, local training documents and interviews with trainers. The report was not published but stored as part of the training documents of the service contract.

We interviewed nine local trainers, at least one local trainer per course, and four master trainers from Wageningen University and KNMI. The *attained* domain relates primarily to students' experiences and learned outcomes from learners perceived by both master trainers and local trainers. The *attained* goals, activities, and assessments were compared with what learners learned by analyzing 38 final case studies submitted at the end of the course as a short presentation in PowerPoint and 73 individual case study proposals submitted at the beginning of the course. Monitoring and evaluation carried out by Leeds University were excluded from this study as these were focused on assessing the learner satisfaction instead of the knowledge and skills achieved.

The analysis was conducted in three phases. First, the collected data was analyzed qualitatively using directed content analysis. In direct content analysis, codes were defined derived from the conceptual framework (Hsieh & Shannon, 2005) on goals, activities, and assessments. Second, potential themes and categories were deduced. In the last phase, all data was organized using a coding tree, including themes and categories. Finally, every code (see Table 3) was typified by the *intended, implemented*, and *attained curriculum* (see Conceptual Framework) (Van Den Akker 2013).

RESULTS

Misalignments in intended goals, activities, and outcomes

The *intended* vision and mission of the ULS was to (1) increase the data uptake from the CDS and (2) ensure the best use of data from the CDS. Indirectly, its mission was to set conditions for getting more CDS data users. However, there were differences in the perception of these *intended* goals as curriculum designers differed in their perception of 'best use of data'. Although some interviewees pointed towards selecting and using the data as 'a way to avoid wrong conclusions best', others highlighted the data selection towards decision-making. These different perceptions implied vague *intended* learning goals without a clear definition of the cognitive levels. We could distinguish between learning for (1) processing and visualization tools, (2) quality and uncertainty in climate data, and (3) climate impacts and risks at the local level for specific sectors and domains. These diverse angles related to different target groups. However, the curriculum designers pointed out that the definition of 'target group' as a data user was controversial and unclear from the beginning. While the tender mentioned a wide range of target groups explicitly as data users (sectorial users, policy-makers, planners, and academic and science communities), curriculum designers reduced this by excluding policy-makers, as they were considered as end-user in comparison with the other target groups. The curriculum designers highlighted that the backgrounds were too different to provide only one training that 'fits them all'. The tender was also pointing to the prerequisite to assess the user's needs to deliver a tailored capacity building. Without a clear understanding of the user's needs and the initial differences in perception, the intended learning goals became ambiguous and, consequently, limiting further specification of the course design.

The *intended* course design aimed at reaching the learning goal through 'blended learning' with various topics and learning activities grounded in adult learning theory. The proposal did not refer to any specific definition of this concept but mentions the use of case studies as a tool to 'capitalize on their own experience' by selecting 'preferred learning situations'. Blended learning was a prerequisite demanded by the ECMWF as a learning approach that could allow professionals to combine their busy agendas with education. The curriculum designers

Table 3 | Code and themes for each curriculum perspective

Code	Theme	Description
Goals	Vision and mission	Describes the desired future position and impact of the service contract through objectives and its approach to reach those objectives
	Target group	The primary group of learners that the course is designed to appeal and usually advertised
Learning activity	Online modules	Content and approach of online lessons from the learning platform
	Training	Content and approach of the C3S Blended Training courses
Assessment	Knowledge and skills	Assessment of methods for measuring knowledge and skills obtained by learners
	Monitoring and evaluation	Formative assessment on the learner's perception of the course

used 'adult learning theory' principles, which became apparent via the inclusion of a case study, to enable adult learners to use their context by bringing their own experiences and applying the knowledge acquired. The case study was a learning activity intended as an individual learning experience to connect context with relevant information. The use of such a case study implies that an understanding of 'best use of data' as data selection towards decision-making was chosen. Yet, the theoretical knowledge offered in the online lessons had a 'best use of data' approach geared at data quality, as this was an ECMWF tender requirement. The 'must-in' topics from the ECMWF tender focused on understanding data and use of CDS such as essential climate variables, seasonal forecasting, climate projections, or CDS Toolbox, among others. Curriculum designers added topics on user engagement and communication in their proposal, although there was no further explanation. Although user engagement and communication were broad and unspecific, they were useful to connect the learner's case study contexts with 'best use of data' for decision-making. However, a direct connection between the offered topics and the related intended target group's needs was missing. The inclusion of abstract concepts like user engagement and communication in the course was a consequence of the ambiguity in the intended goals, target groups, vision, and mission of the C3S Blended Training course.

The *intended* assessment was not a priority with curriculum designers. The proposal mentioned a regular quality assurance through monitoring and evaluation of learner satisfaction after each course instead of assessing knowledge and skills achieved by individual learners. Traditional exams were considered unsuitable because of the course length and the uncertainty on learners' initial knowledge and skills.

Different misalignments were identified within the constructive alignment check on goals, activities, and assessments for the *intended curriculum*. First, the lack of *intended* assessment led to ineffective or less relevant learning activities and goals. Second, the case study as a learning activity was not aligned with the proposed focus on data handling offered in the courses and asked for by the ECMWF tender. Yet, the use of this case study is aligned with the intended mission and vision 'best use of data' towards decision-making.

Misalignments in implemented goals, activities, and outcomes

Before the start of the implementation of the course, a study on training requirements was conducted to be able to better tailor them to the learner's needs. A meta-data analysis and survey between different potential data users was developed to assess the users. The report showed four main points that the C3S Blended courses should consider: (1) the increase of data use is influenced by awareness of what is available and on the understanding of certain concepts; (2) there is a need to better address the complexities around climate services and information provided; (3) personal interest in the use and provision from climate data needs to be addressed; and (4) more attention is needed for how to use the data from the CDS. It is important to consider that the personal interest on the last two points was from the perspective of data users' needs rather than a representation of the intended target group. Although this study guided the design of courses and defined learning goals, activities, and potential outcomes in the implementation phase, the relation with the different target groups remained unclear.

The implemented goals shifted from an increase of data uptake and the best use of data to (1) understanding climate change and impacts, (2) getting acquainted with climate data and terminology, and (3) exploring the CDS potential. The trainers pointed out that this shift in goals came from the juvenile status of the CDS, limiting data availability during the courses, but also as a coping mechanism for the uncertain learner's profile and prior knowledge. The trainers frame this shift as a simplification of the learning goals.

Concerning the learning activities and content of the implemented curriculum, we could distinguish three main learning activities: (1) online lessons, (2) lectures during courses on location, and (3) the case study. The online lessons and lectures focused on understanding climate data (e.g., data resources and discovery, seasonal forecast, models, and reanalysis) and exploring the CDS. A guest lecture was included as a learning activity to contextualize climate impacts. The *intended* content on user engagement and communication was not implemented either in online lessons or lectures. In the case studies, learners would apply the theoretical knowledge acquired through the online lessons and lectures into a real case. Learners could bring in their individual case study proposals, of which trainers selected case studies for groups to work on. The selection was based on sectorial diversity, representation of local interests, and data availability. The case study activities included: theoretical discussions and brainstorms of relevant adaptation issues, potential users, data information needs, and climate pertinent data for the case. The development and actual use of climate data for the case study was not compulsory for the learners. Only learners with programming skills were encouraged to use data. Trainers emphasized the theoretical discussions and brainstorms over the actual use of data.

Local trainers set the final presentation and submission of the case study as a formative assessment to evaluate knowledge and skills achieved by learners. Based on their perception of the implemented course and their evaluation of the case study, trainers also submitted a report to give feedback and suggestions for improvements.

Different misalignments were identified in the implementation phase. First, the case study as a learning activity was still misaligned with the intended goals. While developing a case study on a climate service for adaptation requires wide trans-sectorial and multidisciplinary knowledge, the learning goals focused on the data side. Second, the processes and concepts for designing climate services were not addressed in other learning activities of the course. However, the implementation of the case study as a formative assessment corrected this second intended misalignment.

Attained goals, activities, and outcomes

Local trainers perceived that participants in the trainings attained different learning goals. For example, the local Spanish trainer pointed out that learners with management positions in the private sector aimed to understand climate data and potentially include it in their business. Similarly, a Portuguese trainer pointed out that non-climatic related researchers sought to include climate data in their research. However, trainers observed that, where learners could manage complex data and had coding skills, they lacked theoretical knowledge on climate data. Thus, the *attained* goals became flexible, relevant, and interest-based, tailored to the different categories of learners.

Regarding the learning activities, the individual case study evolved into group case studies. The trainers perceived a vast spectrum of learners from data users to end-users and general-interested learners. The learner's backgrounds varied in sectors, job titles, and prior knowledge on adaptation, climate data, and climate services. Learners with limited experience and prior knowledge could not carry out the case study alone and spontaneously created groups to learn from each other. By learning from this experience and using the learners' heterogeneity, trainers transformed the individual intended case study into a group learning activity.

Local trainers also perceived that groups with different backgrounds were better able to connect climate data with the real-life or policy context than groups with a more homogeneous background. The Portuguese trainer gave a specific 'good' example from a case study on agriculture. He argued that, as the group contained business background learners, they better could address and connect the user needs compared with other groups. Trainers perceived the heterogeneity at the beginning of the courses as challenging to provide a course that 'fits them all'. However, trainers highlighted how their perception changed and saw heterogeneity as an advantage. The heterogeneous groups had more inside knowledge on the case from different perspectives and expertise and experienced more interactive multi-way engagement activities than the homogenous groups.

The case studies were used to assess increases in knowledge and skills, and increases were seen, particularly in the understanding of contextualized climate impacts, climate data, and the CDS. Case studies showed the correct use of scenarios and data set selection from the CDS. Trainers observed a shift from a data-driven initial individual proposal into user-driven case studies after changing to group work in the case studies compared with the first individual proposal with the group case studies. The majority of the first assignments focused on using data and its potential purpose (supply-driven) instead of first identifying the user need (demand-driven). For example, the first assignment adaptation case study (in Italy) was defined as 'Extreme events like heat waves are on the rise in the changing climate scenarios'. Compared with this first assignment, the final adaptation case study included the purpose and use of the information produced.

However, most final case studies did not consider which decisions are being supported by the information provided, resulting in misalignment between the selected user and the information provided. For example, initially in a Portuguese case on 'adaption to climate change in the Tagus River', the learners focused on a better understanding of 'the dynamics of climate change at a transnational river basin scale, and to develop the appropriate elements for an adaptation strategy that would tackle the main water stress risks and potential cross-sectoral interactions'. The first assignment showed the participant rationale: adaptations strategies are directly related to understanding the climate change impacts. The climate service's input was providing information on the climate change impacts on the river basin. But this data orientation missed the decision-maker(s) context for those adaptation strategies and the complexity of stakeholders involved in the same issue. However, the final case study evolved by distinguishing between multiple stakeholders, from politicians in multi-level governance, citizens, NGOs, farmers and agricultural entities, and the energy sector. All stakeholders except those on governance were considered users of the climate services. Although those stakeholders initially were considered having homogeneous interests and power, learners started to recognize the relevance of differentiating between information needs. But the course's supply of climate information was not prepared for this.

Trainers identified three main reasons for this misalignment in the attained courses: the lack of context expertise, the lack of user within the group, and the lack of knowledge and skills provided during the course's learning activities. Trainers argued that without the context-based knowledge from an expert (e.g., hydrologist) and/or the user (e.g., a water manager or policymaker from a municipality), it proved to be challenging to assess the needs and to identify users. Trainers proposed improvements by including learning activities that explicitly link different users with different decisions and, therefore, information. Trainers also suggested including expert's and users within learning group activities.

Although the earning goals were flexibly adjusted in the attainment phase, based on background and interest from the learner, the resulting learning goals were yet covering the required knowledge and skills for climate services in the case study. Therefore, the lack of guidance on knowledge and skills in climate service development could be a relevant factor in the misalignment between user information.

Inconsistencies in curriculum perspectives related to tailored climate services and future implications

The combined results described in the previous results sections in terms of constructive alignment and curriculum perspectives are presented in a matrix (see Table 4).

The learning goals changed or evolved significantly from their intended to implemented to attained forms. The *intended* goals proved overambitious and assumed a one-size-fits-all approach, which turned out not to be feasible for the heterogeneous target group. The implemented and the attained goals became more tailored to the needs and prior knowledge of the different types of learners. Initially, curriculum designers *intended* a slightly different understanding of the 'best use of data' from its direct meaning on improving the supply of reliable and accurate climate information to best information supplied for decision-making. Although this might be perceived as a minor difference, its consequence is significant in the target group. While the 'best use of data' direct meaning suggests that the target groups are actors that directly manipulate data, the 'best' information supplied for decision-making relates to a more complex network of actors. Vincent *et al.* (2018) argue that tailored climate

Table 4 | Matrix of constructive alignment on goals, activities, and assessments

	Learning goals	Activities	Assessments	Misalignments
Intended	Combination 'best use of data' and 'best information supplied for decision-making.' Learning goals: (1) learning processing and visualization, (2) quality and uncertainty in climate data, (3) climate impacts and risks	Online modules on the use of climate data and CDS, communication, and user engagement Individual adaptation case study to design and apply climate data and tools according to context and needs	Not intended	No learning objective related to the development of the case study No assessment Diversity of learners not sufficiently addressed
Implemented	Status CDS, learner uncertainty, and training requirements pushed the shift to general goals: (1) Increase awareness on climate change and its potential impacts (2) Increase knowledge of climate data (3) Explore opportunities for the CDS	Online modules on climate data and use of CDS Expert lecture on local climate change Group work case study: theoretical discussions and brainstorms. Use of data limited to learner's background and data availability	Presentation and submission case study as a formative assessment on which feedback is given	No learning objective related to the development of the case study Online modules and lectures do not offer the knowledge and skills needed to design a climate service for adaptation during the case study
Attained	Implemented goals become flexible, relevant, and interest-based on the learner's profile	Online lessons on climate data, use of CDS, and uncertainty Group case studies with collaboration, interaction, and co-design in more multidisciplinary groups	Presentation and submission of the group case study as a formative assessment on which feedback is given Increase knowledge and skills on the understanding of climate change, climate data, and the CDS. The lack of definition of decisions from users: a misalignment between user information	The lack of learning activities to support the design of a climate service is considered the causes of the misalignment between user information and the lack of decisions
Inconsistencies	Shift due to overambitious 'one-size-fits-all' objective towards flexible learning goals tailored learner's heterogeneity, prior knowledge, and CDS status	Simplification on the development of the case study towards theoretical discussions Learning activities do not adequately prepare for the knowledge and skills needed for collaboration, interaction, and co-design for the case study	The misalignment between user information detected using the case study as the formative assessment	

services require considering a broader spectrum of stakeholders from users and a more comprehensive range of multidisciplinary and cross-sectoral stakeholders. Given this initial ambiguity and uncertainty in the target group, trainers shifted the *implemented* learning goals towards a lower cognitive level set of more general learning goals. However, in the attained curriculum, the ambiguous goals became flexibly adjusted to the target group. The C3S Blended Training courses decided to open itself to this flexibility and embrace it. Trainers highlighted the advantage of opening the course to a heterogeneous target group: because participants learned from each other and experienced more interactive engagement activities and co-production than in the homogenous groups.

The collaborative learning activities among providers and users as implemented by the C3S trainers correspond with calls for more such interactions in literature (Weichselgartner & Kasperson, 2010; Lemos *et al.*, 2012; Vincent *et al.*, 2018; Weichselgartner & Arheimer, 2019). Although improving data availability and better inclusion of the future change in hydrometeorological extremes is essential for climate services, it is insufficient for providing tailored climate services. It requires inclusive and collaborative learning processes between providers and users (Weichselgartner & Kasperson, 2010). Therefore, as the tool to promote the knowledge and skills to achieve inclusive and collaborative learning processes for tailored climate services, capacity building should also explicitly address them. Furthermore, a better understanding of the different stakeholders and users' prior knowledge, experiences, and expectations are essential for more tailored climate services (Weichselgartner & Arheimer, 2019) and, therefore, also crucial for understanding user needs in terms of 'learning needs' (Rauthmann, 2017). Understanding learning needs allows tailored capacity building providing flexible learning goals to the different learning profiles in an inclusive program.

The learning activities not only shifted in content but also in the way the case studies were used in the trainings. The intended learning content aimed for a broad selection of topics, including user engagement and communication. The University of Leeds report on training requirements highlighted the need to address complexities around climate change and climate services. However, the implemented learning activities largely missed that content. But the content and the case study, shifted from individual to group work, driven by the need to cater to the heterogeneity of learners.

The learning activities included discussions and brainstorms. However, the actual use of climate data was limited to learners with programming skills. While learners highly skilled in managing data can benefit from getting acquainted with the CDS and further enhancing their knowledge on climate data, other learners sought the basic understanding of adaptation and climate data for their business. The cases studies promoted collaboration and co-production. Incorporating user engagement processes, including online lessons as an explicit learning goal in capacity building, could be essential to learn the principles of tailored climate services. However, understanding co-production as a type of user engagement is still limited (Alexander & Dessai, 2019). Further research on how co-production relates to climate services could assist in its translation into future capacity building.

Using the case study result for a formative assessment on which feedback was given during the presentation and submission helped produce more user-driven case studies. However, most of the case studies missed the definition of decisions from users, leading to a misalignment between users and information provided.

CONCLUSIONS

In this study, the educational design principles used in the C3S Blended Training courses on climate services were assessed. By evaluating the educational design of the trainings through the lenses of constructive alignment (Biggs, 2003) and the *intended–implemented–attained* curricula (van den Akker 2013), we could analyze how the educational design could be improved to better foster capacity building for tailored climate services. These educational analyses focused on the learning goals, learning activities, and assessments, as well as on the needs of the different learners.

This study shows that capacity building on climate services attracts a heterogeneous background of learners with diverse knowledge and skills. Learners can be roughly divided in service providers, usually with a scientific expert background, and service users, usually operating in policy- and planning-oriented contexts. But within these main groups, many subdivides can be made. Defining standardized learning goals for such a heterogeneous group led to inconsistencies in course design and implementation and consequently to lower then desired attainment levels. Capacity-building design requires understanding the learner's profile to align the goals, activities, and assessments and accommodate for the contribution of different learners in dealing with tailored climate services.

Focusing the learning activities on the use and understanding of climate data does not directly translate into tailored climate services. Learners succeeded in increasing their theoretical knowledge of climatic data. Still, they missed connecting this theoretical knowledge with knowledge needs relevant for specific decision-making contexts. There proved to be a gap between information offered and user information needs. Therefore, capacity building should include content and exercises for designing decision-driven climate service in the goals, activities, and assessment strategies. The case studies, submitted by learners, initially were not meant to close that gap, but under influence of learners seeking collaboration during the trainings and starting to work together in the cases, the case studies largely proved to be a helpful method in closing that gap.

Capacity-building programs designs, like the C3S, are usually not based on educational design principles. This study highlights the usefulness of educational research approaches in the design and implementation of training for climate services for the clear purpose of transitioning to better, tailored, and actionable information. But as climate services themselves aim at conveying information and knowledge and can be regarded as a form of training; the use of educational research approach is expected to be helpful in improving climate service's designs in general. Although this study offers some first ideas to align the educational design, further research is necessary to assess their effectiveness to design tailored climate services.

Collaboration among learners with different backgrounds in the trainings caused a shift from individual learning towards multidisciplinary and cross-sectorial group work in the case studies. This change was not envisaged in the intended learning goals and activities. Collaborative learning should not only be a case study-related activity, but should be included in the content, goals, learning activities, and assessment strategies. Not only in trainings, but also in real-life climate services-boundary crossing among stakeholders and disciplines is required. Therefore, capacity building for climate services should not only focus on providing different stakeholders with relevant knowledge, but also prioritize stimulating different stakeholders to learn and co-create across each other's boundaries. However, literature on climate service still fails to recognize the difficulties associated with organizing collaboration as a design principle for tailored climate services. Yet, more in-depth knowledge is needed to assess to what extent the inclusion of collaborative learning could promote the co-production of tailored climate services within capacity-building programs.

The first-year C3S Blended Training courses reflect the challenges of dealing with a heterogeneous learner group combined with the unbalanced emphasis on supply-driven learning approaches and limited understanding of the principles of tailored climate services. Further research on the nature of tailored climate services and learner profiles for designing capacity-building programs becomes essential. The lack of understanding of these factors impacts the definition of learning goals, teaching and learning activities, and assessment within a capacity-building program (Biggs, 1996). Developing an aligned program in which goals, activities, and assessments are geared towards decision-based and tailored climate service might strongly increase the effectiveness of capacity building for tailored climate services. Those improved climate services might potentially provide actionable knowledge to assist decision-making in adaptation, as it is in the case of the water sector among other sectors in general.

DATA AVAILABILITY STATEMENT

Data cannot be made publicly available; readers should contact the corresponding author for details.

REFERENCES

Alexander, M. & Dessai, S. (2019). What can climate services learn from the broader services literature? *Climatic Change 157*(1), 133–149. doi:10.1007/s10584-019-02388-8.

Biggs, J. (1996). Enhancing teaching through constructive alignment. *Higher Education 32*(3), 347–364.

Biggs, J. B. (2003). Aligning Teaching for Constructing Learning, *The higher educational educational Academy*. Available at: https://www.researchgate.net/publication/255583992_Aligning_Teaching_for_Constructing_Learning (Accessed August 25 2020).

Dilling, L. & Lemos, M. C. (2011). Creating usable science: opportunities and constraints for climate knowledge use and their implications for science policy. *Global Environmental Change 21*(2), 680–689. doi:10.1016/j.gloenvcha.2010.11.006.

Domingos Freires, F. (2016). *World Meteorological Organization Working Together in Weather, Climate and Water Global Framework for Climate Services (GFCS)*. Available at: http://www.wmo.int/pages/gfcs/gfcs_en.html (Accessed December 2 2019).

ECMWF (2017). *Tender Cop_010 User Learning Service*. Available at: https://atmosphere.copernicus.eu/cop010-user-learning-services (Accessed September 13 2020).

ECMWF (no date a). *Copernicus Climate Change Service*. Available at: https://climate.copernicus.eu/ (Accessed January 7 2020).

ECMWF (no date b). *User Learning Services | Copernicus*. Available at: https://climate.copernicus.eu/user-learning-services (Accessed December 2 2019).

García, M. M., Hileman, J. & Bodin, Ö. (2019). Collaboration and conflict in complex water governance systems across a development gradient: addressing common challenges and solutions. *Ecology and Society. Resilience Alliance 24*(3). doi:10.5751/ES-11133-240328.

Hecht, A. D. (1980). Meeting the challenge of climate service in the 1980s. *Bulletin of the American Meteorological Society 65*(4), 365–366.

Hsieh, H. F. & Shannon, S. E. (2005). Three approaches to qualitative content analysis. *Qualitative Health Research* 15(9), 1277–1288. doi:10.1177/1049732305276687.

Karpouzoglou, T., Zulkafli, Z., Grainger, S., Dewulf, A., Buytaert, W. & Hannah, D. M. (2015). Environmental Virtual Observatories (EVOs): prospects for knowledge co-creation and resilience in the information age. *Current Opinion in Environmental Sustainability*, 40–48. doi:10.1016/j.cosust.2015.07.015.

Kirchhoff, C. J., Carmen Lemos, M. & Dessai, S. (2013). Actionable knowledge for environmental decision making: broadening the usability of climate science. *Annual Review of Environment and Resources. Annual Reviews 38*(1), 393–414. doi:10.1146/annurev-environ-022112-112828.

Kuusaana, E. D. & Bukari, K. N. (2015). Land conflicts between smallholders and Fulani pastoralists in Ghana: evidence from the Asante Akim North District (AAND). *Journal of Rural Studies*. doi:10.1016/j.jrurstud.2015.09.009.

Lemos, M. C., Kirchhoff, C. J. & Ramprasad, V. (2012). Narrowing the climate information usability gap. *Nature Climate Change*, 789–794. doi:10.1038/nclimate1614.

Lourenço, T. C., Swart, R., Goosen, H. & Street, R. (2016). The rise of demand-driven climate services. *Nature Climate Change 6*(1), 13–14.

Máñez, M., Zölch, T. & Cortekar, J. (2014). Mapping of Climate Service Providers-Theoretical Foundation and Empirical Results: A German Case Study, CSC Report 15, Climate Service Center, Germany.

Mauser, W., Klepper, G., Rice, M., Schmalzbauer, B. S., Hackmann, H., Leemans, R. & Moore, H. (2013). Transdisciplinary global change research: the co-creation of knowledge for sustainability. *Current Opinion in Environmental Sustainability 5*(3–4), 420–431. doi:10.1016/J.COSUST.2013.07.001.

Nicolescu, B. (1997). The transdisciplinary evolution of the university condition for sustainable development. In *Paper Presented at the International Congress of the International Association of Universities*, 12–14 Nov. 1997, Chulalongkorn University, Bangkok, Thailand.

Rauthmann, J. F. (2017). Situational factors. In *Encyclopedia of Personality and Individual Differences*. Springer International Publishing, pp. 1–4. doi:10.1007/978-3-319-28099-8_1910-1.

Rojas, R., Bennison, G., Galvez, V., Claro, E. & Castelblanco, G. (2020). Advancing collaborative water governance: unravelling stakeholders' relationships and influences in contentious river basins. *Water (Switzerland) 12*(12), 1–25. doi:10.3390/w12123316.

Schuck-Zöller, S., Brinkmann, C. & Rödder, S. (2018). Integrating research and practice in emerging climate services – lessons from other transdisciplinary dialogues. In *Springer Climate*. Springer, pp. 105–118. doi:10.1007/978-3-319-74669-2_8.

Swart, R. J., Bruin, K., de Dhenain, S., Dubois, G., Groot, A. & von der Forst, E. (2017). Developing climate information portals with users: promises and pitfalls. *Climate Services*, 12–22. doi:10.1016/j.cliser.2017.06.008.

van den Akker, J. (2004). Curriculum perspectives: An introduction. In: *Curriculum Landscapes and Trends*. Springer, Dordrecht, pp. 1–10.

van den Akker, J. (2013). Curricular development research as specimen of educational design research. *Educational Design Research* 53–70.

Vaughan, C. & Dessai, S. (2014). Climate services for society: origins, institutional arrangements, and design elements for an evaluation framework. *Wiley Interdisciplinary Reviews: Climate Change 5*(5), 587–603. doi:10.1002/wcc.290.

Vaughan, C., Dessai, S. & Hewitt, C. (2018). 'Surveying climate services: what can we learn from a bird's-eye view? *Weather, Climate, and Society 10*(2), 373–395. doi:10.1175/WCAS-D-17-0030.1.

Vincent, K., Daly, M., Scannell, C. & Leathes, B. (2018). What can climate services learn from theory and practice of co-production? *Climate Services 12*, 48–58. doi:10.1016/j.cliser.2018.11.001.

Weichselgartner, J. & Arheimer, B. (2019). Evolving climate services into knowledge–action systems. *Weather, Climate, and Society 11*(2), 385–399. doi:10.1175/wcas-d-18-0087.1.

Weichselgartner, J. & Kasperson, R. (2010). Barriers in the science-policy-practice interface: toward a knowledge-action-system in global environmental change research. *Global Environmental Change 20*(2), 266–277. doi:10.1016/j.gloenvcha.2009.11.006.

World Meteorological Organization (2010). *Weather, Climate and Water Global Framework for Climate Services (GFCS)*. Available at: http://www.wmo.int/pages/gfcs/gfcs_en.html (Accessed November 1 2019).

World Meteorological Organization (2012). *Guidelines on Frameworks for Climate Services at the National Level World Meteorological Organization, Geneva*.

First received 20 November 2020; accepted in revised form 7 March 2022. Available online 28 March 2022

© 2022 The Authors From Knowledge and Capacity Development to an Implementation Science: doi: 10.2166/wp.2021.071

Governing river rehabilitation projects for transformative capacity development

Patrick Martel [a], Catherine Sutherland [a,*] and Sylvia Hannan [b]

[a] School of Built Environment and Development Studies, University of KwaZulu-Natal, Durban, South Africa
[b] Human Sciences Research Council, Durban, South Africa
*Corresponding author. E-mail: sutherlandc@ukzn.ac.za

PM, 0000-0002-2930-8442; CS, 0000-0002-0415-5232; SH, 0000-0001-8684-7998

ABSTRACT

River rehabilitation projects are framed as water security interventions in South Africa. They aim to address water quality and water quantity issues, as well as to improve socio-ecological relationships. These projects acknowledge the value of capacity building and social learning in enhancing water security. However, they adopt different governance approaches and hence have different knowledge construction and capacity building outcomes. This paper employs a 'governmentality' framework to analyse the capacity development processes within three river rehabilitation projects in Durban, South Africa. The analysis revealed that the three projects, with their different governmentalities, produced different capacity development modalities, which are utilised to sustain 'the object of intervention' in each river rehabilitation project. However, despite these differences, information as the currency of action; the context or site of learning; the importance of building state–citizen relationships; and the need for bridges or intermediaries, emerged as common elements which support capacity building and knowledge sharing across all three projects.

Key words: Capacity building, Governmentality, River rehabilitation projects, Social learning, State–citizen relationships, Water governance

HIGHLIGHTS

- River rehabilitation projects are socio-ecological governance processes.
- Capacity building modalities emerge from different governmentalities of river rehabilitation projects.
- Locally driven river rehabilitation projects emerge through institutional voids.
- Sharing knowledge, understanding socio-ecological relations, state–citizen relationships and intermediaries are critical to transformative river rehabilitation projects.

1. INTRODUCTION

Rivers hold, move, and address multiple urban challenges through their hydrological and ecological processes. In South Africa, these challenges include rapid urbanisation and increasing informality; pollution; poor delivery of services; path dependencies created through hydro-modernist strategies, which use rivers as buffers when engineering systems fail; sand winning; densification and hardening of catchments; poor stormwater management; and cycles of droughts and floods, exacerbated by climate change. If rivers, the 'veins and arteries' of urban systems, are unhealthy and neglected, the entire system is compromised and the quality of life of urban residents, particularly the poor, declines (Turpie *et al.*, 2017a; Anderson *et al.*, 2019). Since 2013, localised river rehabilitation projects, each with their distinct imaginaries, knowledge claims and hydrosocial territories (Boelens *et al.*, 2016; Mills-Novoa *et al.*, 2020) have emerged in Durban, creating territories of intervention (objects) with communities (subjects) to sustain them (Mills-Novoa *et al.*, 2020). These projects have been established by different networks of actors, each with their own governance arrangements and framings of river rehabilitation; and they have created environmental subjects and built capacity in different ways. While they each have their own governmentalities, or rationalities and practices, they all aim to improve water quality and quantity and enhance social learning through the imaginary of river rehabilitation. River rehabilitation projects take on various forms reflecting a wide range of approaches, including technical, socio-technical, and community-based interventions (Keil, 1998; Jewitt *et al.*, 2020). This raises questions about what a rehabilitated river looks like, what the appropriate geographical scale or hydrosocial territory of river rehabilitation should be, and who is responsible for the governing of river rehabilitation (Martel *et al.*, 2022).

Traditionally, river rehabilitation projects have tended to focus on the ecological characteristics of rivers and, therefore, aimed to improve the ecological functioning of degraded ecosystems (King *et al.*, 2003; Grêt-Regamey *et al.*, 2016). The hydrosocial territory of these projects is determined by the spatial demarcation of land-use impacts on rivers at various scales, with the catchment being the most commonly used scale in South Africa. River rehabilitation projects have tended to prioritise technical, scientific, and engineering-based approaches, which do not always consider the socio-economic and political relations that exist within the catchment (Grêt-Regamey *et al.*, 2016; Jewitt *et al.*, 2020). Scientific and technical approaches often render problems technical, de-politicising solutions. This hegemonic imaginary results in the governance context becoming obscured, limiting the imagination and negotiation of alternative paths (Keil, 1998; Moore, 2013; Steelman *et al.*, 2015; Boelens *et al.*, 2016; Hommes *et al.*, 2020). Harrington (2017) states that with the increased range of actors involved in collaborative water governance at both the global and local scale, there is a growing need to consider political and governance processes as being at the core of water issues, including river rehabilitation projects.

This paper compares the capacity building interventions of three river rehabilitation projects established in Durban between 2011 and 2016, using a governmentality framework. The Palmiet Catchment Rehabilitation Project (PCRP), the Aller River Pilot Project (ARPP), and the Sihlanzimvelo Stream Cleaning Programme (SSCP) (see Figure 1) have different imaginaries, logics and rationalities, with capacity development seen as a way of rendering water and people governable in each project. The logics of these projects are self-reinforcing, as the cultivation of subjects helps to sustain the objects of intervention for each project. The flow of knowledge in the capacity development processes is shaped by the initial framings and imaginaries of each project, which impacts the level of transformation achieved (Mills-Novoa *et al.*, 2020). Figure 1 shows the location of the three river rehabilitation projects in Durban.

This paper first outlines the methodology adopted for the study. It then presents the context of Durban's river rehabilitation projects. Next, it explores the meaning of river rehabilitation projects, their imaginaries, and 'what they look like and aim to achieve'. It then defines the hydrosocial territory and analyses the formation of environmental subjects in each case study, revealing the different governmentalities that were established. Finally, it presents the different modalities of capacity building within river rehabilitation projects that emerged in Durban, identifying four characteristics of capacity building in projects that aim to be transformative.

2. METHODOLOGY

The paper draws on research conducted from 2014 to 2020 on emergent river rehabilitation projects in Durban. The authors have been directly involved in or conducted research on all three projects, and therefore acknowledge the impact of their positionality in shaping their understanding and analysis. The research is based on a wide range of primary data collected through interviews, focus groups, and learning laboratories on governance

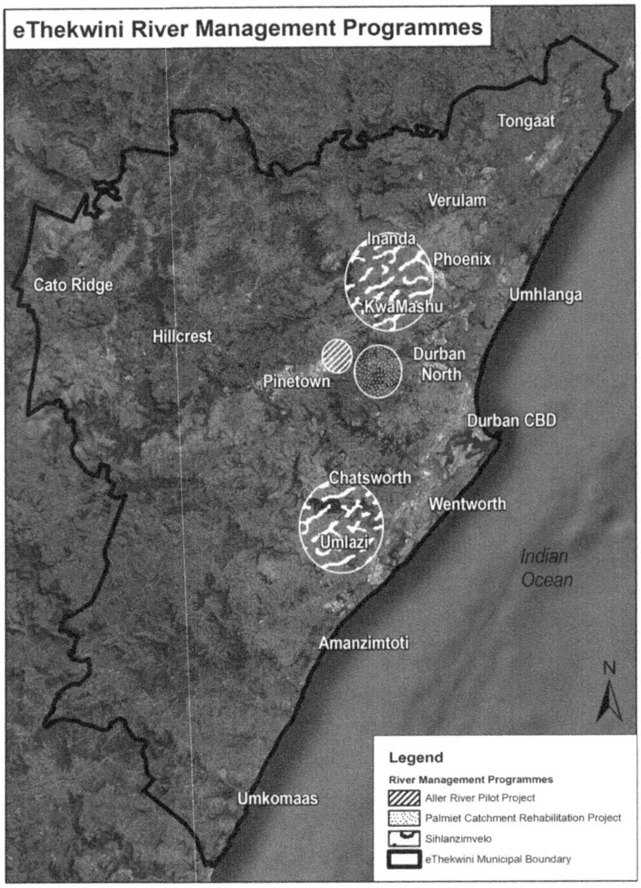

Fig. 1. | The location of the three river rehabilitation projects in Durban.

approaches and capacity building in river rehabilitation projects, through various water governance research programmes the authors have been a part of (SANCOOP Climways, WRC 2354, GCRF U-RES, LIRA 2030[1]). In this paper, primary data and project-related data (grey literature) on governance approaches and capacity building in the three case studies is analysed using a thematic approach, identifying each project's actors, hydrosocial territory, imaginary or rationality, discourses and practices. From this, the rationalities of government and technologies of rule for each case study are identified. Foucault's concept of governmentality

---

[1] SANCOOP CLIMWAYS, funded by the National Research Foundation, South Africa, and Research Council of Norway, title: *Climate Change and Urban Water Governance (2014–2017). Pathways to Social Transformation* which focused on river rehabilitation projects; Water Research Commission (WRC) 2354, title: *Enhancing Water Security through Restoration and Maintenance of Ecological Infrastructure: Global Lessons from the uMngeni Catchment, South Africa (2014–2020)*, which focused on investing in ecological infrastructure, including in social learning to enhance water security; Global Change Research Fund Resilience Foundation Awards, title: *Foundations for Climate Resilient and Sustainable Growing Settlements (U-RES), (2017–2019)*, which focused on building resilience in the Palmiet Catchment; LIRA 2030 Africa, title: *Transforming South African Cities in a Changing Climate, 2018–2020*, which focused on understanding the role and governance of river rehabilitation projects in Durban.

was employed to analyse how different forms of governing result in different forms of knowledge and capacity building outcomes for the partners in these projects, including state, civil society organisations, citizens, and research institutions.

3. RIVER REHABILITATION PROJECTS IN THE DURBAN CONTEXT

River rehabilitation projects in South Africa are classified as water security interventions, as they aim to address both water quality and quantity concerns, which impact on human and environmental well-being (Jewitt *et al.*, 2020). These projects aim to promote change, including to invest in ecological infrastructure, promote behaviour change, improve waste management, reduce flood risk, address water pollution, remove invasive alien plant species, support socio-economic development and poverty alleviation and build social cohesion through partnerships between the local state, civil society organisations, and research institutions (Martel *et al.*, 2022; Jewitt *et al.*, 2020, C40 Cities Finance Facility, 2021). Both globally, and in South Africa, the protection and enhancement of ecological infrastructure, defined as 'ecosystems that deliver services to society, functioning as a nature-based equivalent of, or complement to, built infrastructure' (Cumming *et al.*, 2017: 53), is recognised as being essential to river rehabilitation projects (Daily & Matson, 2008; Jewitt *et al.*, 2020). Jewitt *et al.* (2020) expanded the definition of ecological infrastructure, recognising that ecosystems both produce and deliver services. Protecting and rehabilitating biodiversity and ecosystem services is, therefore, a main aim of these programmes. In the global South, particularly in Africa, the protection and enhancement of ecological infrastructure cannot be addressed in isolation. It needs to be integrated into poverty alleviation and development programmes, ensuring co-benefits for humans and nature (Roberts *et al.*, 2012).

Under South African legislation, river catchments and their sub-catchments fall under Water Management Areas (WMAs) with their associated Catchment Management Agencies (CMAs). The National Water Act (No. 36 of 1998) requires the establishment of CMAs to govern water in South Africa. CMAs require engagement between state and non-state actors and have a strong focus on the biophysical health of catchments, as well as social transformation, supporting the principles of integrated water resources management (Meissner *et al.*, 2017). Within this institutional imaginary, each CMA is required to compile a Catchment Management Strategy and ensure water resources planning. A catchment management strategy must set principles for allocating water to existing and prospective users, taking into account all matters relevant to the protection, use, development, conservation, management, and control of water resources (Republic of South Africa, 1998). In the Breede-Gouritz Water Management Area's Catchment Management Strategy, improving river health is considered a main priority, with the vision of this strategy, or its imaginary stated as 'Healthy water resources, for all, forever' (Breede-Gouritz Catchment Management Agency, 2017). However, seven of the nine CMAs in South Africa, including the Pongola-Mzimkulu WMA, within which Durban is located, do not have a functional or influential CMA in place, due to human resource and financial constraints and a lack of political will to support decentralised water management (Meissner *et al.*, 2017).

At a regional scale, the uMngeni Ecological Infrastructure Partnership (UEIP), established in November 2013, promotes the participation of multiple organisations, including national, provincial and local government departments, business and academic institutions, as well as civil society, in water governance in the uMngeni Catchment. This catchment, located in the Pongola-Mzimkulu WMA, KwaZulu-Natal, supports over six million people, including the residents of Durban, in South Africa's third largest regional economy (Hay, 2017; Jewitt *et al.*, 2020). Given the absence of a functioning CMA, the UEIP focuses on integrating investments in ecological infrastructure, built infrastructure and social learning, to address water security.

3.1. Durban's river rehabilitation efforts

Durban, or the eThekwini Municipal Area (EMA), is located on the east coast of South Africa, in the province of KwaZulu-Natal. It is the country's third largest city, with a population of 3.7 million people. The city experiences high levels of poverty and inequality, with 42% of residents considered poor, and a quarter (26%) living in informal settlements (eThekwini Municipality, 2016/2017). The EMA is located within one of 36 global biodiversity hotspots, the Maputaland Pondoland Albany biodiversity hotspot. As a result, it contains a wealth of natural capital, much of which is located within the Durban Metropolitan Open Space System (DMOSS). The environmental assets of the city include 18 major river catchments and 16 estuaries (Turpie *et al.*, 2017b; see Figure 2).

In terms of water resources management, eThekwini Municipality falls within the dysfunctional Pongola-Mzimkulu WMA. This has created an institutional void, as a government structure designed to govern water and support the implementation of legislation and policy, is not active. In response to both this governance vacuum, and concern about the quality of rivers, a diverse range of river rehabilitation projects have emerged within the municipality. These projects have adopted different governance approaches, each with their own imaginaries, hydrosocial territories, knowledge construction processes and capacity building approaches and outcomes.

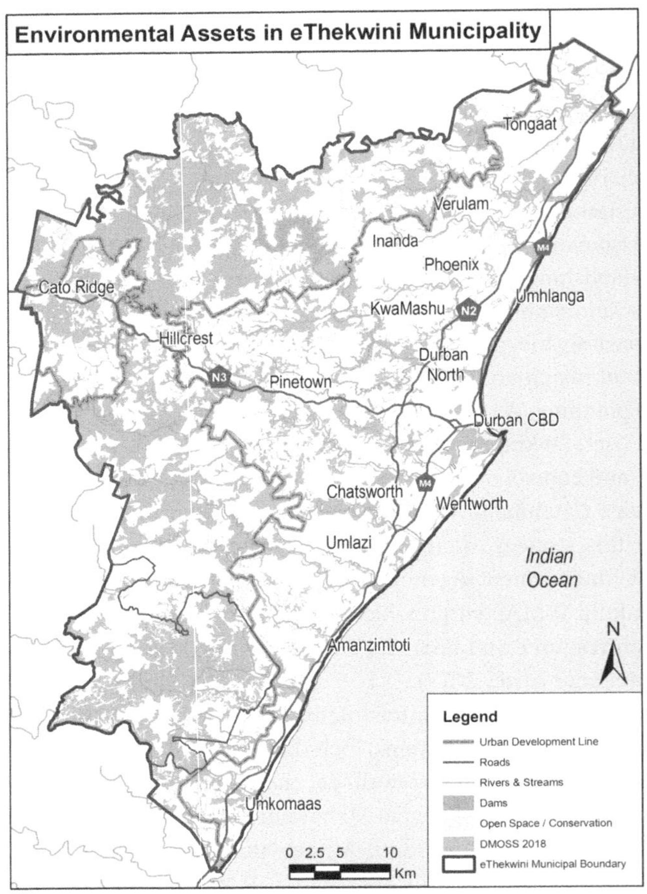

Fig. 2. | Durban's environmental assets.

The C40 Cities Network (2017), in examining community-based interventions to improve river health in Durban, highlighted the importance of cultivating active and engaged citizens who play a vital role in mobilising local action and contributing towards broader resilience objectives. Capacity building includes awareness creation, knowledge building and skills development, which enables citizens to find employment and to contribute towards a resilient society and economy (C40 Cities Network, 2017). Capacity development includes activities such as education and training, joint research, as well as knowledge networking and partnerships.

Durban's urban resilience efforts, through the leadership of the Environmental Planning and Climate Protection Department, support the development of an informed and educated citizenry for transformative sustainable development. In 2013, Durban was selected to participate in the Rockefeller Foundation's 100 Resilient Cities (100RC) Programme (Roberts *et al.*, 2020). A key output from this programme was the development of a 'Resilience Strategy' for the city. Through a range of engagements, city stakeholders identified resilience focus areas. This included the importance of the natural environment as the foundation for sustainable development, human well-being, and resilience. The need to develop active, educated and engaged citizens was acknowledged (C40 Cities Network, 2017; Sutherland *et al.*, 2019a).

Drawing on the goals and outcomes of Durban's Resilience Strategy (eThekwini Municipality, 2017), a number of municipal departments, including the Environmental Planning and Climate Protection Department, eThekwini Water and Sanitation Unit, Coastal Stormwater and Catchment Management Department, and the newly established Sustainable and Resilient City Initiatives Unit, increased their focus on capacity building and social learning for environmental citizenship. This focus on community engagement and the co-production of knowledge, along with the emergence of collaborative river rehabilitation projects, created new platforms of governance. These platforms began to intersect with the efforts of the C40 Finance Facility to support climate adaptation in the city, through the leadership and visioning of the Coastal Stormwater and Catchment Management Department and the Climate Protection Branch. The imaginary of river rehabilitation, with its hydrosocial territory of the river or the catchment, was rationalised as a means to address multiple environmental and social challenges in the city. This led to the development of Durban's Transformative Riverine Management Programme in 2019, with the first phase being the development of the business case for riverine management (C40, 2021). This programme has now become the frame for Durban's multiple river rehabilitation projects, its climate adaptation efforts in catchments, and its river-based socio-ecological poverty alleviation programmes.

In this paper, we conceptualise river rehabilitation projects as 'improvement schemes' (Li, 2007), and endeavour to understand what these projects aim to change, and what calculated means are applied to improve conditions through capacity development – by highlighting which actors are targeting different objects and subjects. The local actors framing the river rehabilitation projects (their imaginaries and hydrosocial territories) and implementing particular interventions and activities, include local government officials, civil society organisations, university-based researchers, and in one case, consultants. Despite their different aims and objectives, all three projects have involved the state and local community members, capacitating these groups to varying degrees, in order to satisfy project specific requirements.

4. CREATING ENVIRONMENTAL SUBJECTS: THE GOVERNMENTALITY OF CAPACITY BUILDING

Governmentality, developed by Foucault (1977–1978), focuses on the art of governing. This notion analyses the organised practices through which citizens are governed and self-govern, and how they are produced as subjects of the state or those doing the governing, in order to fulfil policy (Foucault, 1991). Governmentality, therefore, examines how the conduct of humans is shaped by calculated means to achieve certain ends (Li, 2007). With a focus on governmentality, this paper argues that capacity development is a crucial consideration, especially when people are used as a calculated means to achieve a certain end. Different governmentalities, therefore, shape the outcomes of capacity development as they frame the training, skills, and practices learned, as well as how social learning takes place. In our work, we use two governmentality terms for analysis. First, political rationalities, defined as a specific form of reasoning, which outlines the telos, or ultimate aim of action, or the means to achieve it (Lemke, 2002). Secondly, technologies of rule, which are the particular tools and tactics used to govern (Li, 2007). Hydrosocial territories, as a conceptual framework, aligns with governmentality theory, as it 'grounds and bounds the object created through the regime of practices' surrounding river rehabilitation projects (Mills-Novoa *et al.*, 2020: 91).

Governmentality explores the 'everyday relations of power extending beyond the central state, … that order societies and make human beings into subjects, i.e. subjects to government by others and to self-government' (Merlingen, 2003: 366). Governmentality can be indirect and unseen, shaping political decision-making and social behaviour, by influencing the generation of expert knowledge, regulatory practices and societal expectations (Foucault, 1991). Power is exercised through the rationalities or mentalities of government and technical means or 'technologies of rule', which are the policies and practices of those in power, who do the governing (Merlingen, 2003; Li, 2007; Dean, 2010).

Li (2007) identifies two main practices required to transform 'the will to improve' into precise programmes. First is the practice of problematisation, which identifies and frames the challenges that need to be resolved. Second is the practice of rendering problems technical, through a range of practices associated with characterising the domain to be governed (Li, 2007). Political rationalities are an element of government in their own right and produce a discursive field in which exercising power in a particular way and framing targets of government, such as river rehabilitation, is considered to be rational (Lemke, 2002; Djama *et al.*, 2011). Technologies of rule signify the particular tools and tactics used to govern (Djama *et al.*, 2011) and represent the organised, structured, controlled, and reproduced modes of power that follow a specific rationality when governing (Lemke, 2002), which in this case are capacity building approaches. These are 'the actual mechanisms through which authorities of various sorts have sought to shape, normalise and instrumentalise the conduct, thought, decisions, and aspirations of others [the governed] in order to achieve the objectives they [authorities/governors] consider desirable' (Miller & Rose, 1990: 8).

In order to govern the environment, three strategies are required (Agrawal, 2005). These are the creation of governmentalized localities, able to be regulated in specified domains (catchments); the opening of spaces (territorial and administrative) (river rehabilitation projects) in which new regulatory communities can function; and the production of environmental subjects, defined as 'people who have come to think and act in new ways in relation to the environmental domain being governed …' (citizens) (Agrawal, 2005: 7). The making of environmental subjects is based on the different forms of regulation and the practices through which people get involved in them (Agrawal, 2005), with the subject and the environment being simultaneously remade and redefined in particular contexts, leading to new technologies of government (Agrawal, 2005).

River rehabilitation projects create environmental subjects in various ways, including through capacity building or development initiatives. Eade (2007) emphasises that capacity development cannot be reduced to a universal blueprint. Rather an engaged outsider who understands the local context should 'support the capacity of local people in determining their own values and priorities, to organise themselves to act upon and sustain these for the common good, and to shape the moral and physical universe that we all share' (Eade, 2007: 632). The United Nations Development Program (UNDP, 2011: iv), drawing on research undertaken at IHE Delft, define capacity development for environmental sustainability as 'the process by which individuals, organisations and societies strengthen their ability to address environmental issues, manage natural resource issues, and mainstream environmental sustainability into development policies, plans and decisions'. Central to capacity development for environmental sustainability is the complex interrelationship between the environment and socio-economic development, and the relationships between the actors involved in these relations (Ziervogel *et al.*, 2021).

Capacity development modalities, being programmatic in character, represent a form of technologies of rule. The capacity development activities are conceptualised by project actors as a means to solve their various water-related challenges or to facilitate some form of change. The understanding and application of capacity development has changed over time, from primarily focusing on training, to a wider range of modalities, including education and training, joint research and co-production of knowledge, advisory services, e-learning and distance learning, and knowledge networking and partnerships (Wehn de Montalivo & Alaerts, 2013; Wehn *et al.*, 2015). Although the principle of including stakeholders in river management has become widely accepted, consensus has not been reached on the specific goals or forms of stakeholder involvement and the nature of capacity building (Buchecker *et al.*, 2013). Despite this lack of consensus, these processes are recognised as promoting co-ownership of the process design and providing opportunities for learning between stakeholders.

5. THE GOVERNMENTALITIES AND CAPACITY BUILDING OUTCOMES OF DURBAN'S RIVER REHABILITATION PROJECTS

The three case studies represent a range of water-related interventions with the same imaginaries (river rehabilitation for social and environmental benefits), but different rationalities, hydrosocial territories, and technologies of rule (forms of capacity building). The 'government', or those doing the governing, varies across the projects.

5.1. The governmentalities of Durban's river rehabilitation projects

5.1.1. Palmiet Catchment Rehabilitation Project

The Palmiet Catchment is a sub-catchment of the uMngeni Catchment, which was selected as eThekwini Municipality's proof-of-concept case study in the UEIP for assessing the value of investing in ecological infrastructure in securing water (Vogel *et al.*, 2016; Jewitt *et al.*, 2020). In 2014, local government officials from the Environmental Planning and Climate Protection Department, together with researchers from the School of Built Environment and Development Studies, University of KwaZulu-Natal (UKZN), and civil society organisation Palmiet River Watch, as well as community members from Quarry Road West informal settlement, established the PCRP. The aim of the PCRP is to rehabilitate the Palmiet River and its catchment, to improve water and climate governance, and to provide evidence of the value of ecological infrastructure in securing water (Martel & Sutherland, 2019). The PCRP comprises of a voluntary network of actors who form the PCRP's Community of Innovation that guides the project. The collective object of intervention in the PCRP is the Palmiet Catchment, and being reflective of self-government, the 'loose' environmental subjects are those who co-produced the Action Plan. Given the informal and voluntary nature of this project, more formal arrangements are associated with small-scale initiatives – which target specific sites or communities and are based on the respective project actor's interest and aspiration. The premise is that actors voluntarily align their initiatives with the PCRP's co-produced Action Plan.

Prior to 2019, the project was funded through international and national research programmes led by UKZN researchers[2] and Durban's Research Action Partnership (DRAP), which includes municipal funding. The PCRP has evolved into an innovative governance platform that supports a wide range of local government-led and community-based initiatives to support the resilience and sustainability of the Palmiet Catchment (Sutherland *et al.*, 2019a, 2019b). The PCRP has shown its resilience as a governance platform by remaining functional since 2014, attracting new funding and projects over time. The most significant of these is the investment by the Development Bank of Southern Africa in a four-year project (2019–2022) managed by eThekwini Municipality and

[2] SANCOOP CLIMWAYS, Water Research Commission Project 2354, GCRF U-RES, City of Bremen-GIZ, NRF Community of Innovation, University of Manchester's SCI Urban Data Justice Project.

Table 1 | The governmentality of the Palmiet Catchment Rehabilitation Project (PCRP).

	PCRP
The 'government' or 'authority' (who does the governing)	Collection of actors: UEIP, state, research institutions, and civil society. Primary actors who lead the programme are from the university, eThekwini Municipality and River Watch and they form the Community of Innovation.
Problematisation	River rehabilitation requires rehabilitation of the catchment, treating 'the cause and not just the symptoms' through an understanding of and change in socio-ecological relationships across the entire catchment.
Political rationalities	Understand socio-economic, political, and environmental relations in the catchment to achieve transformative change. River rehabilitation needs to address both formality and informality in the catchment. River rehabilitation needs to emerge through the co-production of knowledge and engagement by multiple actors to produce what a 'rehabilitated river looks like and is' in each particular context.
Technologies of rule	Multiple actors produced an action plan for catchment rehabilitation through co-engagement and recognition of different forms of knowledge. Emergent projects, which arise through co-engagement and context-specific challenges, form the focus of catchment rehabilitation projects.
Spatial scale (hydrosocial territory)	Palmiet Catchment
Environmental subjects	All actors engaged in the PCRP, and communities living within the Palmiet Catchment.
Social learning and capacity building	Capacity building of multiple actors through co-production of knowledge and co-engaged learning around socio-ecological relations and transformative catchment management.

implemented by consultants, GroundTruth, that serves as a proof of concept for investment in ecological infrastructure. While the PCRP has remained in place, its governance platform has come under threat as a result of conflicts between different actors, who frame river rehabilitation in different ways. The most significant tension arises around the acceptance and management of informality in the catchment and the failure of the municipality to address water pollution and service-related issues. One of the main challenges is that the Climate Protection Branch is not an implementing department, but rather informs and facilitates implementation by departments such as the eThekwini Water and Sanitation Unit and the Coastal Stormwater and Catchment Management Department. This has led to dissatisfaction from residents in the catchment around the issue of water quality and service provision, which is yet to improve, resulting in mistrust of the outcomes and benefits of the PCRP. Protests by communities in the informal settlement and letters of complaint sent to city officials are evidence of this lack of faith in transformative change. City officials and researchers from UKZN have highlighted how challenging it is to improve water quality across a catchment over the short term (7 years), arguing that progress has been made, but that real change takes a long time, particularly in a city facing multiple challenges and competing priorities. Table 1 outlines the governmentality elements structuring the PCRP.

5.1.2. Aller River Pilot Project

The Aller River is located in a sub-catchment adjacent to the Palmiet Catchment and flows into the lower uMngeni River. eThekwini Municipality funded several pilot projects as part of the development of its Resilience Strategy, including the ARPP, which focused on river rehabilitation and water governance along a section of the

Aller River (Martel & Sutherland, 2019). The ARPP was initiated by a civil society organisation, the eThekwini Conservancies Forum (ECF), in 2016, as part of their broader Take Back Our Rivers Programme. Conservancies are voluntary conservation organisations, which link communities, state departments and conservation efforts, to promote stewardship of natural resources and achieve broader socio-economic benefits at a community level. The project aimed to develop and implement community interventions to restore the health of a 5.6 km section of the Aller River, which is characterised by poor water quality (Martel & Sutherland, 2019). This stretch of river, as well as the local community impacting on it, represents the object of intervention. The ARPP harnessed the knowledge and resources of the conservancy movement to develop a community-based approach to river rehabilitation. It was envisaged that the ARPP would contribute to improved stewardship of eThekwini Municipality's water resources; enhance livelihood opportunities in low-income communities; and strengthen resilience in the face of climate change (Martel *et al.*, 2017). The project succeeded in building state–citizen relationships around pollution reporting and river management, and led to environmental education activities within local schools. There is a strong focus on behaviour change, where selected community members have been employed as Eco-Champs (community-based champions) in order to engage with local community members – with the goal of promoting river health and community ownership and responsibility. However, the impact of the project was limited due to a lack of sustained funding and the need for Conservancy members to continue to manage the project, which had not become self-sustaining, both in terms of leadership and funding. Conflict between stakeholders was less evident, as the ECF designed the programme and implemented it, with limited space to debate the problematisation and objectives of the ARPP. The governmentality elements structuring the ARPP are summarised in Table 2.

Table 2 | The governmentality of the Aller River Pilot Project (ARRP).

	ARRP
The 'government' or 'authority' (who does the governing)	eThekwini Conservancies Forum (civil society organisation).
Problematisation	River pollution is a result of failed relationships, and a lack of education and accountability of actors responsible for river pollution and rehabilitation: namely community, state, industry.
Political rationalities	Improve river health through community responsibility and ownership. Cultivate local influencers to create environmental awareness; build relationships between communities, industry, and the state.
Technologies of rule	Community-based eco-champs identified, appointed, and trained: build their capacity to influence and engage with other actors. Eco-champs engage with the local community and municipal officials around pollution issues. Community awareness raising and education by eco-champs. River walks (monitoring).
Spatial scale (hydrosocial territory)	Riparian zone and communities adjacent to riparian zone along 5.6 km of Aller River.
Environmental subjects	Eco-champs and by association community members in communities adjacent to the 5.6 km of the Aller River.
Social learning and capacity building	Capacity building of eco-champs through knowledge transfer from the Duzi UMngeni Conservation Trust (DUCT), which is then transferred on to community adjacent to Aller River and local schools.

5.1.3. Sihlanzimvelo Stream Cleaning Programme

The SCCP was implemented in 2011 by the eThekwini Municipality Roads and Stormwater Maintenance Department in consultation with City Councillors. The programme aims to remove litter/waste and invasive alien plant species from waterways to reduce stormwater blockages and create employment. With a predominantly operational focus, the initial design of the programme focused on improving cost efficiencies in service delivery. Within the programme, community co-operatives, funded by the local state, employ local community members to clean 300 km of stream banks and culverts, removing solid waste and invasive alien plant species. The streams are located in high density, low-income settlements, where poor river quality has negative impacts on human health, and there are risks of flooding. A consultant is responsible for managing the project, including appointing community assessors to monitor implementation and build local awareness (C40 Cities Finance Facility, 2017). The programme has been successful in building state–citizen relations around service delivery and has been used as an exemplary river rehabilitation programme upon which the business case for financing river rehabilitation in eThekwini Municipality is being built (C40 Cities Finance Facility, 2021). The programme predominantly focuses on cleaning and clearing, with the co-benefits being employment for local communities and the raising of environmental awareness around the value of rivers and their catchments. As a technical stream cleaning programme, which includes training and awareness raising of community members employed through the co-operatives, the project has been a success. It has cleaned up rivers and stormwater systems and has created jobs for local community members in partnership with the municipality, which has been a political goal for poor urban communities for many years. However, it remains a technical and operational programme with limited input from communities living adjacent to rivers and stormwater systems about their relationships with these systems. Table 3 highlights the governmentality elements structuring the SSCP.

Table 3 | The governmentality of the Sihlanzimvelo Stream Cleaning Programme.

	Sihlanzimvelo
The 'government' or 'authority' (who does the governing)	eThekwini Municipality (with support from consultants).
Problematisation	Dense settlements adjacent to river courses are poorly serviced and experience poor waste management which results in poor environmental management of riverine corridors. This results in damage to urban infrastructure through flooding and build-up of solid and ecological waste.
Political rationalities	Maintenance of waterways through alien invasive clearing and removal of pollution prevents damage to built infrastructure (culverts) during flooding. Job creation through investment in ecological infrastructure. Cost effective means to secure infrastructure.
Technologies of rule	Technical and operational interventions undertaken by community co-operatives, overseen by consultants, with support from municipality led SSCP. Clearing of alien invasive species, collection and removal of solid waste and erosion prevention interventions undertaken by community co-operatives. Training of community co-operatives to undertake rehabilitation tasks, provision of protective clothing and tools.
Spatial scale (hydrosocial territory)	Watercourses (5 km stretches) identified by the SSCP project manager.
Environmental subjects	Working members of the co-operatives, stream cleaners, and by association community members.
Social learning and capacity building	Capacity building of co-operatives around stream clearing, business development, and health and safety, who work in local spaces on rehabilitation of water courses.

Drawing on Mills-Novoa *et al.* (2020), each project has constructed a territory (the object) that is in need of intervention and has used capacity building as a means to target certain populations (subjects), both internal and external groups, in order to achieve project goals. The governmentalities or ontologies of rule (Rose & Miller, 2010) evident in each case study (Tables 1–3) produced different modalities of capacity building. The form of governing, problematisation and type of environmental subjects created, shaped the capacity building outcomes, which in turn reproduced the modes of governance, relations of power and the type of environmental subjects best suited to fulfil the rationalities or policies of the government. This resonates with Li (2007) concept of trusteeship, where those who govern, develop the capacity of others and direct it in benevolent ways. This power is often exercised at a distance, structuring the field of possible actions and modifying processes in ways that appear as ordinary or part of everyday life. In this way the desires and actions of these powerful actors are not overt nor immediately evident and are not perceived as impositions (Li, 2007).

5.2. The modalities of capacity building in river rehabilitation projects in Durban

Three modalities of capacity building, or technologies of rule, which produce particular environmental subjects, have emerged through the process of governing river rehabilitation projects. These are creating community-based champions, adopting a developmental social justice approach, and producing environmental subjects through community-based stream clearing co-operatives (see Figure 3).

5.2.1. Creating community-based champions

The creation of community-based champions in the ARPP (environmental subjects), using the eco-champs model developed by DUCT, reflects the first modality of capacity building. These community-based champions are agents of change. They are influencers and bridges, promoting positive environmental behaviour and ownership in targeted communities, and connecting citizens to the local state. At the outset of the programme, environmental

Fig. 3. | Creating environmental subjects through different modalities of capacity building.

knowledge was transferred from DUCT and the ECF to environmental champions (eco-champs) through capacity development in the form of training. This established a frame within which the eco-champs thought about the environment and took action for its protection, in relation to the built environment and social systems. During the capacity development training, the eco-champs were encouraged to engage and reflect on their own experiences in relation to how environmental challenges along the river were being framed, thereby connecting their imaginaries and knowledge with those of the ECF and the rationalities of government of the 'Take-Back-Our-Rivers' programme. This ensured that knowledge produced reflected the realities on the ground and encouraged social learning.

The building of relationships and partnerships was prioritised, as these champions interacted with community members, as well as with local government officials in departments related to river pollution. One of the tactics used in this modality of capacity building was the collection, generation, and transfer of information. This included the use of mobile phones or other GPS enabled devices; the population of spreadsheets (capturing details on surcharging manholes, response times, causes of blockages); citizen science (miniSASS); and visual observations during river walks. This information enabled the eco-champs to participate in broader governance networks, particularly with local government officials. The data were also used as evidence of their local environmental expertise when they acted as bridges between the community and local government. The eco-champs became the first group of environmental subjects produced through the project. They then imparted their knowledge and used their data to produce environmental subjects in both their community and in the local state, requiring positive responses from both sets of actors to achieve the goal of river rehabilitation. The hydrosocial territory of this project, which was a zone adjacent to 5.6 km of the Aller River, contained and created boundaries for the knowledge production that took place between those doing the governing, both the ECF and the environmental subjects the project created.

5.2.2. A developmental social justice approach

Environmental subjects can be created through a developmental, social justice model which is not hierarchical, but rather horizontal in nature. The PCRP is framed by network governance, prioritising participatory governance, the co-production of knowledge, and transdisciplinarity. Within the governance arena of this project, a diverse range of actors learn from each other, in understanding risks and opportunities in the catchment, in the context of formality and informality.

Different kinds of environmental subjects are produced and transformed by this network of actors, through the issues and projects that emerge in the PCRP. For example, the Palmiet River Watch invokes environmental subjects who engage in citizen science and the reporting of environmental pollution, focusing on the ecology of the river. Researchers from UKZN encourage the creation of environmental subjects within Quarry Road West informal settlement, who act as a voice and as agents of change for informal communities, with a focus on shifting socio-ecological relations. The state invokes environmental subjects through the Action Plan of the PCRP and short-term river clean ups, using a model similar to that used in the SSCP.

The capacity development methods employed include joint research activities and the co-production of knowledge, reflecting that all actors in the PCRP have become environmental subjects, who are benefiting from shared learning. Tacit and experiential community-based knowledge is moved through the networks of the PCRP's governance arena, using co-engagement, presentations in meetings, social learning, dialogue, and publications to inform policy makers within the municipality. This approach has been instrumental in building state–citizen relationships, with university researchers and Palmiet River Watch acting as intermediaries (Sutherland *et al.*, 2019b). This co-produced knowledge has moved into other government programmes, including Durban's Resilience Strategy, iQhaza Lethu Informal Settlement Upgrading Programme, and Durban's Transformative Riverine

Management Programme, through the efforts of officials, academics and community members, and has supported local government's understanding of the politics and challenges of rehabilitating a catchment with multiple land uses. Given the complexity of the hydrosocial territory of this project, namely the catchment with its multiple land uses and challenges, the boundaries of the project and what it seeks to achieve shift and are broadened as issues in the catchment arise. Similar outcomes of the benefits of locally produced knowledge are evident in Ziervogel *et al.* (2021)'s action research undertaken in Cape Town on climate adaptation programmes.

5.2.3. Environmental subjects within community co-operatives

Environmental subjects have been cultivated through eThekwini Municipality's Sihlanzimvelo[3] Stream Cleaning Programme, which has an infrastructure maintenance and poverty alleviation focus. Working within community co-operatives, these environmental subjects have an operational focus, and hence their capacity development modality reflects a more practical orientation, aligned with a technical and managerial form of governance. Prior to the implementation of Sihlanzimvelo, the concept was presented to local councillors and ward committees, who helped to shape the tactics of government used in the project. This includes the identification of the most polluted water courses, the selection of co-operatives, the level of payment for services and the form of training, including environmental and business training, and the provision of tools and safety equipment. The model uses money as the incentive for change, as co-operatives and their members do not get paid if they do not achieve their stream clearing tasks and activities.

Within the SCPP, the collection of data is completed by consultant assessors, who convey this information to the project managers. Project decision-making and calculation lie at the project manager level. This means that community co-operative members are not responsible for knowledge construction processes, and hence are not required to develop capacity in knowledge production. Community co-operative members are provided with basic business and administrative skills. This ensures that the co-operatives are managed as businesses and have the capability (and required paperwork) to re-apply as community co-operatives every 3 years. This is in line with the technical, managerial, and hierarchical form of governance evident in this project. The hydrosocial territory of each co-operative is defined as a local area requiring a clean-up, which has been identified by local councillors in conjunction with municipal officials, and so the boundaries of the territory are clearly defined, which makes project implementation focused and efficient.

While the governmentalities of each river rehabilitation project have produced different capacity building approaches and outcomes, four main themes have emerged as being central to all three projects.

5.3. Central and critical capacity building elements

Four common themes of capacity building were evident across the three projects: information as the currency of action; the site of learning; the importance of building state–citizen relationships; and the need for bridges or intermediaries (see Figure 4).

Information is used as the value add or currency for non-state actors engaging in governance networks. Here, the collection and use of information, and context specific, local knowledge, has empowered citizens to engage with the state, and other actors. In some cases, this information and knowledge has been produced within predetermined frames (community-based eco-champ modality), while in other cases, communities themselves have constructed what knowledge or information matters and should be included (developmental social justice modality). In the co-operative modality, the information and knowledge are framed by what is required by the operational practices of the project. The governmentality of each project shapes the level of openness or control

[3] When translated from Zulu into English, Sihlanzimvelo means 'we care for our environment'.

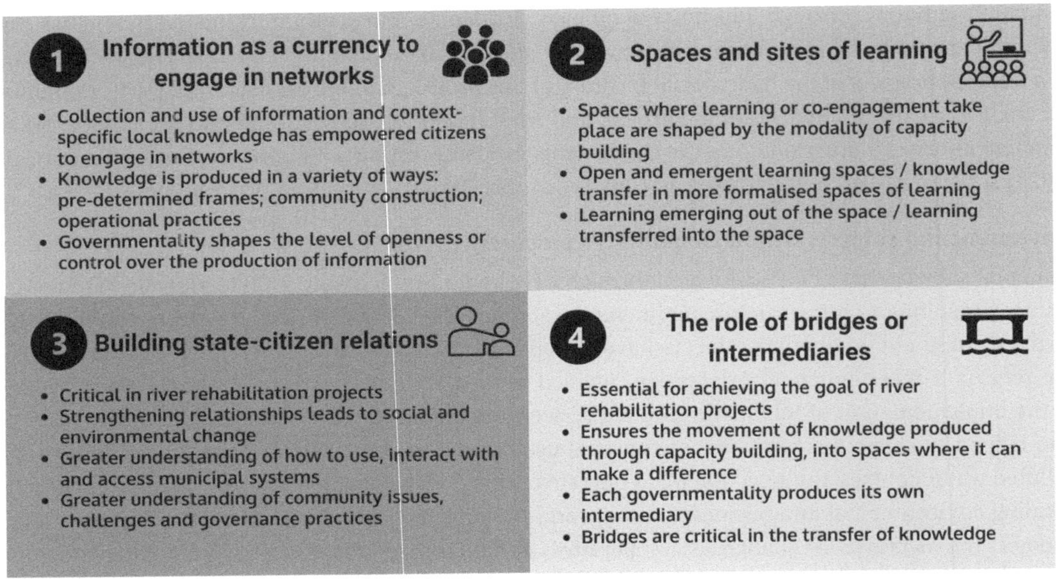

Fig. 4. | Critical elements of capacity building programmes.

over the production of information and determines where this knowledge can flow. Building capacity in people enables local actors to exercise power to govern themselves, their communities, and the local state, as was evident in the ARPP and the PCRP.

The second theme focuses on spaces and sites of learning. The spaces where the learning or co-engagement takes place are shaped by the modality of capacity building, as co-produced modes result in an open and emergent learning space, while the transfer of knowledge creates a more formalised space of learning. In all three projects, the site of learning and capacity building is focused on the river catchment itself. In some cases, the knowledge and learning emerge from this physical space, for example in the PCRP; while, in the other two cases, learning is transferred into the space. In the case of the PCRP, the form of learning is more transformative as it enables co-engaged learning pathways and produces learning environments that are more participatory, reflexive, and learner-led (O'Donoghue *et al.*, 2018). However, this is more challenging, as environmental subjects are not produced through calculated means in the form that those managing river rehabilitation projects may require to ensure their goals and objectives are met, which can result in conflict between project actors. However, the benefit of this approach is that all actors are included in the institutionalisation of river rehabilitation projects, as practical engagement, inclusive participation, critical review, re-visioning, and the deliberation of the way forward is achieved (O'Donoghue *et al.*, 2018). In all projects, the cultivation of environmental subjects is of benefit to the state, as departments within local government have access to more grounded information about built infrastructure, the local social and political system, local environmental knowledge, and practices within catchments. This creates the potential for the state to fulfil its mandate by having access to this information.

Thirdly, capacity building in all three cases has resulted in the building of relationships between citizens and the state, which is considered critical in river rehabilitation projects. The power relations between citizens and the state vary in all three models, with the developmental social justice model producing a more horizontal and less vertical set of relations, due to the co-production of knowledge that takes place. However, in each case, state–citizen relationships are strengthened, and this leads to the potential for, and realisation of, social and environmental change. The environmental subjects also develop a deeper understanding of how they can use, interact with, and access municipal systems.

Finally, bridges or intermediaries have emerged as essential to achieving the goals of river rehabilitation projects and ensuring that the knowledge produced through capacity building is transferred into spaces where it can begin to make a difference. The presence of, and need for, intermediaries reveals that capacity can be built in individuals, but bridges become important in the transfer of knowledge between communities and local government (Sutherland *et al.*, 2019b).

6. CONCLUSION

Rivers are a nexus of socio-economic, political, and environmental relations that constitute a city. They receive, move, and process the outcomes of urban life, and in turn they shape and transform the environments they flow through. The relations between society and rivers are mediated by hard engineered and ecological infrastructure, and human action, which can have both positive and negative effects. The improvement of relationships between rivers and society needs to be built through adaptive, emergent, interdependent and dynamic interventions, rather than the technical, linear, and deterministic interventions of the past (Ziervogel *et al.*, 2021). This paper has presented the different governmentalities that have emerged in river rehabilitation projects in Durban at the local scale, as the actors doing the governing (a network of actors; state, university and civil society; a civil society organisation and the state) have produced their own problematisation, rationalities of government and technologies of rule for each project within particular hydrosocial territories. In the absence of the overarching CMA imaginary and its technologies of rule, each project develops its own tactics aligned with its rationalities, which creates different environmental subjects, and different outcomes.

Capacity building and social learning are identified as a technology of rule, as a means of shaping the behaviour and action of those living adjacent to rivers. Each governmentality produces its own form of capacity building, ranging from the direct transfer and uptake of technical knowledge to a more open and transformative approach of knowledge building from below. This shapes the outcomes of the river rehabilitation projects as it creates the frame within which social change can take place, which sustains each project's specific imaginary, based on their object of intervention. Four main themes emerged as being central to capacity building. Knowledge and information, and the way it flows and where it flows to, is critical in all projects. Information is the currency of exchange, however, the movement of this knowledge through the data information chain to policy makers, will determine the extent of transformative change (Sutherland *et al.*, 2019b; Ziervogel, *et al.*, 2021). The site of learning and its openness determines whose knowledge matters most in shaping river rehabilitation outcomes. Building state–citizen relations through the capacity building of all actors is a cornerstone to transformative river rehabilitation projects. Finally, the role of bridges or intermediaries is essential, while state–citizen relations are being developed and secured.

River rehabilitation projects are framed in different ways, with technical and ecological approaches being broadened to include social, political, and economic dimensions. By including a wide range of environmental subjects in social learning processes, which are inclusive and participatory, those with the will and power to promote change, can begin to pursue the much-needed goal of rehabilitating catchments and rivers. The challenges of urbanism in the global South means that this is a challenging task that cannot be easily achieved in the short term; however, new modes of governance and capacity building are emerging, which have the potential to support transformation and sustainability over the longer term.

ACKNOWLEDGEMENTS

The authors would like to thank the multiple stakeholders involved in Durban's River Rehabilitation Projects for sharing their knowledge and insights with us. They would also like to thank the Durban Research Action Partnership (University of KwaZulu-Natal and eThekwini Municipality) for funding this research. The anonymous reviewers who provided valuable comments on this paper are thanked for their input and insights. Michela du Sart, from EduAction, is acknowledged for producing the maps in this paper.

DATA AVAILABILITY STATEMENT

Data cannot be made publicly available; readers should contact the corresponding author for details.

REFERENCES

Agrawal, A. (2005). *The Politics of Nature and the Making of Environmental Subjects Environmentality: Technologies of Government and the Making of Subjects*. Duke University Press, Durham.

Anderson, E. P., Jackson, S., Tharme, R. E., Douglas, M., Flotemersch, J. E., Zwarteveen, M., Lokgariwar, C., Montoya, M., Wali, A., Tipa, G. T., Jardine, T. D., Olden, J. D., Cheng, L., Conallin, J., Cosens, B., Dickens, C., Garrick, D., Groenfeldt, D., Kabogo, J., Roux, D. J., Ruhi, A. & Arthington, A. H. (2019). Understanding rivers and their social relations: a critical step to advance environmental water management. *WIRES Water 6*(6), 1–21.

Boelens, R., Hoogesteger, J., Swyngedouw, E., Vos, J. & Wester, P. (2016). Hydrosocial territories: a political ecology perspective. *Water International 41*(1), 1–14.

Breede-Gouritz Catchment Management Agency (2017). *Catchment Management Strategy for the Breede-Gouritz Water Management Area*. Breede-Overberg Catchment Management Agency, Worcester, South Africa.

Buchecker, M., Salvini, G., Di Baldassarre, G., Semenzi, E., Maidl, E. & Marcomini, A. (2013). The role of risk perception in making flood risk management more effective. *Natural Hazards and Earth System Sciences 13*, 3013–3030.

C40 Cities Finance Facility (2021). *Business Case for Durban's Transformative Riverine Management Programme*. Report Produced by FutureWorks for the C40 Cities Finance Facility.

C40 Cities Network (2017). *Case Study 100 Resilient Cities Pilot Project: Community Based Interventions to Improve River Health (Aller River)*. C40 Finance Facility Network, Durban. Available at: https://www.c40.org/case_studies/community-based-interventions-to-improve-river-health.

Cumming, T. L., Shackleton, R. T., Forster, J., Dini, J., Khan, A., Gumula, M. & Kubiszewski, I. (2017). Achieving the national development agenda and the Sustainable Development Goals (SDGs) through investment in ecological infrastructure: a case study of South Africa. *Ecosystem Services 27*(Part B), 253–260.

Daily, G. C. & Matson, P. A. (2008). Ecosystem services: from theory to implementation. *Proceedings of the National Academy of Sciences of the United States of America 105*, 9455–9456.

Dean, M. (2010). *Governmentality: Power and Rule in Modern Society*, Second Edition. SAGE Publications Ltd, Thousand Oaks.

Djama, M., Fouilleux, E. & Vagneron, I. (2011). Standard-setting, certifying and benchmarking: a governmentality approach to sustainability standards in the agro-food sector. In: *Governing Through Standards: Origins, Drivers and Limitations*. Ponte, S., Gibbon, P. & Vestergaard, J. (eds.). Palgrave Macmillan, London, UK.

Eade, D. (2007). Capacity building: who builds whose capacity? *Development in Practice 17*(4/5), 630–639.

eThekwini Municipality (2016/2017). *Spatial Development Framework*. eThekwini Municipality, Durban.

eThekwini Municipality (2017). *Durban's Resilience Strategy*. eThekwini Municipality, Durban.

Foucault, M. (1991). Governmentality. In: *The Foucault Effect: Studies in Governmentality*. Burchell, G., Gordon, C. & Miller, P. (eds.). Harvester Wheatsheaf, Hemel Hempstead, pp. 87–104.

Grêt-Regamey, A., Weibel, B., Vollmer, D., Burlando, P. & Girot, C. (2016). River rehabilitation as an opportunity for ecological landscape design. *Sustainable Cities and Society 20*, 142–146.

Harrington, C. (2017). The political ontology of collaborative water governance. *Water International 42*(3), 254–270.

Hay, D. (2017). *Our Water Our Future: Securing the Water Resources of the uMngeni River Basin.* Institute of Natural Resources, Pietermaritzburg, South Africa.
Hommes, L., Boelens, R., Bleeker, S., Stoltenborg, D., Duarte-Abadia, B. & Vos, J. (2020). Water governmentalities: the shaping of hydrosocial territories, water transfers and rural-urban subjects in Latin America. *Environment and Planning E: Nature and Space* 3(2), 399–402.
Jewitt, G. P. W., Sutherland, C., Browne, M., Stuart-Hill, S., Risko, S., Martel, P., Taylor, J. & Varghese, M. (2020). *Enhancing Water Security Through Restoration and Maintenance of Ecological Infrastructure: Lessons from the UMngeni River Catchment, South Africa. Vol 1: Final Report.* Report Submitted to the Water Research Commission. WRC Report No. TT 815/20.
Keil, R. (1998). Greening the polis or policing ecology? Local environmental politics and urban civil society in Los Angeles. In: *Cities for Citizens: Planning and the Rise of Civil Society in a Global Age.* Douglass, M. & Friedmann, J. (eds.). John Wiley and Sons, Chichester.
King, J. M., Scheepers, A., Fisher, R. C., Reinecke, M. K. & Smith, L. B. (2003). *River Rehabilitation: Literature Review, Case Studies and Emerging Principles.* Report to the Water Research Commission, WRC 1161.
Lemke, T. (2002). Foucault, governmentality, and critique. *Rethinking Marxism 14*(3), 49–64.
Li, T. M. (2007). *The Will to Improve: Governmentality, Development, and the Practice of Politics.* Duke University Press, Durham, NC.
Martel, P. & Sutherland, C. (2019). Governing river rehabilitation for climate adaptation and water security in Durban, South Africa. In: *The Geography of Climate Change Adaptation in Urban Africa.* Cobbinah, P. B. & Addaney, M. (eds.). Palgrave MacMillan, London.
Martel, P., Sutherland, C., Buthelezi, S. & Khumalo, D. (2017). *Aller River Pilot Project External Evaluation. First External Evaluation.* Unpublished Report Prepared for eThekwini Conservancies Forum, University of KwaZulu-Natal, Durban.
Martel, P., Sutherland, C., Hannan, S. & Magwaza, F. (2022). River rehabilitation projects, and their associated ecological infrastructure, as a spatial expression of a more resilient and sustainable Durban. In: *Sustainable Urban Futures in Africa.* Cobbinah, P. & Addaney, M. (eds.). Routledge, Oxfordshire.
Meissner, R., Stuart-Hill, S. & Nakhooda, Z. (2017). The establishment of catchment management agencies in South Africa with reference to the Flussgebietsgemeinschaft Elbe: some practical considerations. In: *Freshwater Governance for the 21st Century.* Karar, E. (ed.). Springer, Dordrecht.
Merlingen, M. (2003). Governmentality: towards a Foucauldian framework for the study of IGOs. *Cooperation and Conflict: Journal of the Nordic International Studies Association 38*(4), 361–384.
Miller, P. & Rose, N. (1990). Governing economic life. *Economy and Society 19*(1), 1–31.
Mills-Novoa, M., Boelens, R., Hoogesteger, J. & Vos, J. (2020). Governmentalities, hydrosocial territories & recognition politics: the making of objects and subjects for climate change adaptation in Ecuador. *Geoforum 115*, 90–101.
Moore, M. L. (2013). Perspectives of complexity in water governance: local experiences of global trends. *Water Alternatives* 6(3), 487–505.
O'Donoghue, R., Taylor, J. & Venter, V. (2018). How are learning and training environments transforming with education for sustainable development? In: *Issues and Trends in Education for Sustainable Development.* Leicht, A., Heiss, J. & Byun, W. (eds.). UNESCO, Paris. Available at: http://unesdoc.unesco.org/images/0026/002614/261445E.pdf.
Republic of South Africa (1998). National Water Act, Act 36 of 1998. Government Printers, Pretoria.
Roberts, D., Boon, R., Diederichs, N., Douwes, E., Govender, N., McInnes, A., Maclean, C., O'Donoghue, S. & Spires, M. (2012). Exploring ecosystem-based adaptation in Durban, South Africa: 'Learning-by-Doing' at the local government coal face. *Environment and Urbanization 24*(1), 167–195.
Roberts, D., Douwes, J., Sutherland, C. & Sim, V. (2020). Durban's 100 resilient cities journey: governing resilience from within. *Environment and Urbanisation 32*(2), 547–568.
Rose, N. & Miller, P. (2010). Political power beyond the state: problematics of government. *British Journal of Sociology 61*(1), 271–303.
Steelman, T., Nichols, E. G., James, A., Bradford, L., Ebersöhn, L., Scherman, V., Omidire, F., Bunn, D. N., Twine, W. & McHale, M. (2015). Practicing the science of sustainability: the challenges of transdisciplinarity in a developing world context'. *Sustainability Science 10*, 581–599.
Sutherland, C., Roberts, D. & Douwes, J. (2019a). Constructing resilience at three different scales: the 100 resilient cities programme, Durban's resilience journey and everyday resilience in the Palmiet rehabilitation. *Human Geography 12*(1), 33–49.

Sutherland, C., Mazeka, B., Buthelezi, S., Khumalo, D. & Martel, P. (2019b). Making Informal Settlements 'Visible' Through Datafication: A Case Study of Quarry Road West Informal Settlement, Durban, South Africa (September 26, 2019). Development Informatics Working Paper No. 83. Available at http://dx.doi.org/10.2139/ssrn.3460252.

Turpie, J., Letley, G., Chyrstal, R., Corbella, S. & Stretch, D. (2017a). *Promoting Green Urban Development in Africa: Enhancing the Relationship Between Urbanization, Environmental Assets and Ecosystem Services. Part 2: Evaluating the Potential Returns to Investing in Green Urban Development in Durban.* Report Prepared by Anchor Environmental Consultants for AECOM on Behalf of the World Bank.

Turpie, J., Letley, G., Chyrstal, R., Corbella, S. & Stretch, D. (2017b). *Promoting Green Urban Development in Africa: Enhancing the Relationship Between Urbanization, Environmental Assets and Ecosystem Services. Part 1: A Spatial Valuation of the Natural and Semi-Natural Open Space Areas in EThekwini Municipality.* Report Prepared by Anchor Environmental Consultants for AECOM on Behalf of the World Bank.

United Nations Development Program (UNDP) (2011). *Practitioner's Guide: Capacity Development for Environmental Sustainability.* Available at: https://sustainabledevelopment.un.org/content/documents/954017_UNDP_Practitioner_Guide_CD%20for%20Sustainability.pdf.

Vogel, C., Scott, D., Culwick, C. & Sutherland, C. (2016). Environmental problem solving in South Africa: harnessing creative imaginaries to address 'wicked' challenges and opportunities. *South African Geographical Journal 98*(3), 515–530.

Wehn, U., Irvine, K., Jaspers, F., Douven, W., Pathirana, A. & de Ruyter, E. (2015). Strengthening water governance in the global south: role and international experiences of UNESCO-IHE in capacity development. *Water Governance 5*, 26–34.

Wehn de Montalivo, U. & Alaerts, G. (2013). Leadership in knowledge and capacity development in the water sector: a status review. *Water Policy 15*, 1–14.

Ziervogel, G., Enqvist, J., Metelerkamp, L. & van Breda, J. (2021). Supporting transformative climate adaptation: community-level capacity building and knowledge co-creation in South Africa. *Climate Policy 21*, 1–16.

First received 27 March 2021; accepted in revised form 1 October 2021. Available online 18 October 2021

© 2022 The Authors From Knowledge and Capacity Development to an Implementation Science: doi: 10.2166/wp.2022.229

Capacity development for the Bangladesh Delta Plan from the perspective of delta professionals: A qualitative study

Ashraful Kabir[a,b], Abu Syed [a], Chris Zevenbergen [c], Jannatul Ferdous[a] and Assela Pathirana [c,*]

[a] Bangladesh Centre for Advanced Studies, Dhaka, Bangladesh
[b] School of Public Health and Preventive Medicine, Monash University, Melbourne, Australia
[c] IHE Delft Institute for Water Education, Delft, The Netherlands
*Corresponding author. E-mail: a.pathirana@un-ihe.org

AS, 0000-0003-4256-6793; AP, 0000-0003-0907-1764

ABSTRACT

Many deltaic regions around the world are exploring or adopting planning initiatives to address the complex challenges associated with future economic development and climate change. A key feature is their adaptive nature, also coined as adaptive delta management (ADM). ADM refers to the development of strategies and measures that allow us to adapt to change over a long time. The Government of Bangladesh has recently approved the Bangladesh Delta Plan 2100 (BDP2100) – a long-term adaptive strategy for sustainable water resources management. This study aims to explore the future capacity requirements from the perspective of water/delta professionals for implementing BDP2100. This calls for structural and institutional changes, and new skills and competencies to address the specific organisational and individual needs for effective delivery. A capacity development programme as an integral part of BDP2100 is currently under development. Adopting a qualitative approach, 16 key informant interviews were conducted with experts and professionals from the water and water-related sectors in Bangladesh. A thematic analysis approach was used to analyse the data. The participants reflected that an integrated and adaptive system is critical for sustainable delta management; collaboration and coordination between water/delta professionals (e.g., policy planners, consultants, and academics) and delta expert stakeholders with deep local familiarity (e.g., communities, activists, and civil society members) were deemed essential. In addition, a powerful independent institutional body is recommended to provide stewardship and leadership in the implementation process.

Key words: Adaptive delta management, Bangladesh Delta Plan 2100, Capacity development, Climate change, Water resources management

HIGHLIGHTS

- Deltas have embraced adaptive delta management (ADM).
- BDP2100 – ADM blueprint for Bangladesh.
- Needs quantitative and qualitative changes in water sector capacity.
- Investigated capacity needs for the success of BDP2100 as perceived by sector experts.
- Integrated, adaptive approach, and interdisciplinary collaboration should be central in capacity building (CB).
- Independent institutional body should coordinate the sector CB needs.

INTRODUCTION

Adaptation has been recognised as key to address uncertainties associated with climate change (Henstra, 2016; Zandvoort *et al.*, 2017). In the past years, many deltaic regions worldwide adopted adaptation mechanisms for addressing long-term climate-induced vulnerabilities and uncertainties (Swanson & Bhadwal, 2009; Jeuken *et al.*, 2014; Nicholls *et al.*, 2018; Zegwaard *et al.*, 2019). Literature presents several adaptive approaches to support long-term planning and flexibility, acknowledging uncertainties in key drivers of change (Walker *et al.*, 2001; Lempert *et al.*, 2003; Pahl-Wostl, 2007; Haasnoot *et al.*, 2013). Particularly, adaptive delta management (ADM) is increasingly applied to deal with uncertainties in climate change and socio-economic projection scenarios in the design and planning phases of flood risk management strategies (Zevenbergen *et al.*, 2018). ADM refers to the development of strategies and measures which are able to adapt to changes occurring over a long period (Marchand & Ludwig, 2014; Dewulf & Termeer, 2015; Zevenbergen *et al.*, 2018; Kulsum *et al.*, 2021). The ADM approach became a priority for the deltas transforming rapidly due to sea-level rise and other consequences of climate change, and to the direct impact of the climate change drivers: rapid population increase, urbanisation, deforestation, etc. (Huq & Ayer, 2008). Humans have transformed many deltas in the world to a so-called Anthropocene state (Renaud *et al.*, 2013) which are pushing them towards new tipping points that may threaten the livelihood of the communities living in the deltas as well as the sustainability and resilience of their environment (Mintzer, 1988; Giosan *et al.*, 2014).

Bangladesh is covered mainly by the Bengal delta, the largest delta on earth. Global climate change and regional (Ganges-Brahmaputra basin) and local changes continue to affect the ecosystems and livelihoods of the communities of the country (Ali, 1999; Karim & Mimura, 2008; Shahid, 2011; Nicholls *et al.*, 2016). Bangladesh ranks as the seventh most affected country by climate-related impacts and associated vulnerabilities (Eckstein *et al.*, 2021). These vulnerabilities can be exacerbated by many factors such as disadvantageous geographic location, topography, rapidly expanding urban sprawl, high poverty levels, population density, and poor institutional capacity (Goosen *et al.*, 2018; Khan *et al.*, 2021).

In response to the climate-related impacts and associated vulnerabilities, the Bangladesh Delta Plan 2100 (BDP2100) was formulated as a long-term comprehensive strategic and planning initiative to secure sustainable economic growth, enhance environmental conservation, strengthen resilience to natural disasters, and promote effective water governance (Bangladesh Planning Commission, 2018). In September 2018, the Government of Bangladesh (GoB) endorsed BDP2100. ADM is at the heart of BDP2100 taking into account multiple socio-economic and climate change scenarios allowing to change course when the future unfolds differently than anticipated (Timmermans *et al.*, 2015). Hence, it focuses neither on the business-as-usual nor the worst-case scenario but on multiple adaptation pathways that embrace the reality of uncertainty, and sometimes unpredictability, of future states (Zevenbergen *et al.*, 2018). Implementing ADM requires a continuous intensive interaction between different levels of government and a governance system relying on monitoring and evaluation. When new information becomes available – in the planning and programming and in the implementation phases – the decision to change course needs to be based on an institutionally embedded rationale, and coordinated. Often, the central government coordinates the development of the strategy. The coordination of the programming and the execution of measures are typically carried out by regional and local authorities. Especially in situations where large, expensive interventions are required for improving the functioning of the system as a whole, and the cost in terms of negative consequences to local and regional interests is high, it is important that all government levels are committed to the same outcome. ADM has been first applied in the programming of the Dutch Delta Plan and soon after in a few other deltas (Wiens *et al.*, 2017; Radhakrishnan *et al.*, 2018). ADM cannot be transposed easily from one country to another at the policy formulation level as it demands a fundamental change in institutional capacity at multiple levels, including new knowledge and skills, hierarchical and operational relationships, and policy frameworks, and, hence, it depends on local socio-economic characteristics, culture, and governance (Zevenbergen *et al.*, 2013; Hasan *et al.*, 2020).

Evidently, effective implementation of BDP2100 requires significant capacity development at different levels. The implementation of a large programme like BDP2100 generates different dynamics than its formulation. From a top-down perspective, needs have been identified as key to the implementation: for integrating and mainstreaming the BDP2100 programme, for coordinating and collaborating among the agencies and stakeholders, for translating the Plan into sectoral short-term plans, for building the institutional framework, for mobilising resources and prioritising programme interventions, for conducting result-based monitoring and evaluation, and for creating and translating evidence-based knowledge into practices (de Heer, 2020). As such, the capacity for improved water development is viewed as a holistic and complex mechanism that functions at individual, institutional/organisational, and contextual levels. These levels are inter-related and influence each other. At the individual level, skills and competencies are to be developed; then, these are exchanged and shared among peers and groups to adjust them and confirm their suitability at the organisational level, and eventually are applied in an enabling environment (Blokland *et al.*, 2009). The Dutch Delta Approach has been referred to as a successful guidance frame for the Bangladesh BDP2100 (Hasan *et al.*, 2020). The premise of the study presented in this paper is that improved capacities at different government levels are therefore essential for further successful replication and adjustment during implementation. The study is based on interviews with experienced water professionals currently working 'on the ground' in Bangladesh in the water domain; it explores capacity requirements in water management skills and institutions to render the implementation of the BDP2100 possible.

METHODS AND STUDY DESIGN

An exploratory qualitative study was applied. We conducted 16 Key Informant Interviews (KIIs) with respondents holding complementary knowledge regarding water resource management. We selected participants based on their engagement in the implementation process of BDP2100 and their representativity for the Bangladesh water sector. Considering the constraints in resources, time, logistics (e.g., partly due to lockdowns it was difficult to identify and approach participants located outside Dhaka), and the availability of participants, this study stage focuses on participants from inside the domain of water and water-related activities. Interviews from a wider group of delta professionals (such as policy planners and analysts, consultants, academics, water resources economists, lawyers, and environmentalists) and delta stakeholders with deep familiarity with living in, and managing deltas (such as community representatives, environmental activists, and civil society members) are likely to add further insights at a later stage.

Thus, we first identified potential participants and sought their support, clarifying the study's aims, objectives, and expectations. They were selected based on: (i) at least 15 years of professional experience in relevant water and water-related sectors, (ii) their availability, and (iii) achieving an overall broad representation of the sector (government agencies, independent and non-governmental research and policy bodies, and universities, that are engaged in project design and implementation, policy formulation and planning, communication and advocacy, teaching, and consultancy) (Table 1). The number of interviews was based on the principle of data saturation, i.e., ending the interviews when no new information, dimension or concept emerged, as proposed by Guest *et al.* (2006). A stepwise procedure was followed to determine data saturation. Multiple independent interviewers conducted interviews concurrently and finalised the initial/axial codes of the harvested information after completing the first pilot interviews. After having conducted two-thirds of the interview, the research team independently developed the codes and checked whether no new information or themes emerged. If the team

Table 1 | Profile of the participants included in the key informant interviews (n=16).

Interview/ Participant No	Organisation/Institution	Areas of expertise	Gender
P1	Institute of Water Modeling Bangladesh (IWMDB)	30 years of experience in mathematical hydraulic modelling, planning of water and water-related studies	M
P2	Bangladesh Water Development Board (BWDB)	20 years of experience in surface water and groundwater management	M
P3	Bangladesh Centre for Advanced Studies (BCAS)	35 years of experience in teaching and research in water resources, management, and capacity development	M
P4	Bangladesh Water Development Board (BWDB)	25 years of experience in designing and executing surface water and groundwater management programmes	M
P5	Bangladesh University of Engineering and Technology (BUET)	20 years of experience in teaching and research in water resource engineering, management, and policy planning	F
P6	Centre for Environmental and Geographic Information Services (CEGIS)	23 years of experience in project design and implementation in system development and water capacity development	M
P7	Department of Agriculture Extension (DoAE)	34 years of experience in project design and implementation of sustainable agriculture capacity development	M
P8	Bangladesh Centre for Advanced Studies (BCAS)	19 years of project design and research experience in natural resource management, sustainable agriculture, and urban planning	M
P9	Bangladesh Centre for Advanced Studies (BCAS)	22 years of project teaching and research experience in water resources, environmental sustainability, and capacity development	M
P10	Water Resources Planning Organization (WARPO)	18 years of project management experience in the area of water resource, management, and adaptation	M
P11	Local Government Engineering Department (LGED)	23 years of project management experience in the area of sustainable water resource management	M
P12	Local Government Engineering Department (LGED)	16 years of project management experience in sustainable surface water resource management	M
P13	Local Government Engineering Department (LGED)	17 years of experience in policy planning and implementation in sustainable water resource management	M
P14	Water Resources Planning Organization (WARPO)	24 years of experience in project management and implementation in water resource management	M
P15	Institute of Water Modeling Bangladesh (IWMDB)	17 years of experience in hydraulic modelling	M
P16	Institute of Water Modeling Bangladesh (IWMDB)	26 years of experience in hydraulic modelling and water-related policy planning	M

reached agreement that this was the case, data saturation was considered achieved. Importantly, this was an iterative process shuttling back and forth between data collection, interpretation, and coding, during which the team members also reflected on their own roles as interviewers to minimise bias. The discussions followed a semi-structured question guideline that explored the changes perceived necessary at the individual level in the water professional, in educational institutions, and in service delivery and regulatory agencies. The question guide itself was piloted and enhanced beforehand.

The medium of conversation was Bangla – the native language of both interviewers and interviewees. The interviews were audio-recorded with permission of the interviewees. A team of two interviewers (social-science graduates) conducted the interviews, while a research assistant (engineering graduate) took field notes. The interviewers were trained and experienced in the qualitative approach and data collection techniques. Before starting the interview, the research team built a good rapport with the interviewees and discussed the purpose of the study. Each interview took on average 30–45 min.

Data were analysed using thematic analysis (Braun & Clarke, 2006) to separate out distinct concepts and statements and recognise recurrence in a rigorous process. First, we transcribed all interviews verbatim from the audio records and translated the transcripts into English. Then, we developed several codes – meaningful information, ideas, or dimensions – able to capture key considerations relevant to the study objectives. A hybrid approach combining deductive and inductive coding was applied to identify the themes related to the implementation of BDP2100. Initially, a list of codes was developed to align with information available in the literature and the understanding and experiences gathered during the interviewees' engagement in their respective work domains. Three researchers repeatedly went through the transcription line-by-line and coded in line with the predefined set of codes (Fereday & Muir-Cochrane, 2006). Concurrently, additional codes were identified to fit the relevant information or statements observed during the reading of the transcriptions. The final codebook was developed by combining these. A few clusters were analysed based on the similarities of the identified codes. In the final stage, several themes were merged to report the findings. The independent coders maintained formal reviews to resolve any disagreement.

RESULTS

Five themes emerged: (1) Perceptions on potential delta professionals and stakeholders, (2) Required knowledge, skills, and attitudes, (3) Experienced vs. young delta professionals, (4) Organisational culture and capacity, and (5) Educational capacity. Within these themes, 13 sub-themes emerged (Table 2).

(1) *Potential delta professionals and stakeholders, as perceived by the interviewees*

The perceived delta professionals and 'expert stakeholders' are defined as those who are likely to have the potential to effectively contribute to the implementation of the BDP2100 through their specific skills and knowledge, and day-to-day local life experience. Over three-quarters (13 out of 16) of the participants perceived that the potential delta professionals and stakeholders would likely contribute to design and/or implementation of projects/interventions to develop infrastructure facilities, build organisational capacities, facilitate research initiatives, and enhance capacity relevant to BDP2100. The participants stated that BDP2100 should be considered an integrated, holistic effort. Therefore, professionals with various disciplines and perspectives such as engineering, social sciences, fisheries, environmental sciences, forestry, economics, livestock husbandry, and agriculture need to be included in the implementation process. Engaging professionals from diverse disciplines and backgrounds will create a more comprehensive understanding of the delta dynamic, enabling them to design and execute effective interventions in the delta. One participant stated '[The] Delta Plan is a holistic approach. It should engage experts and professionals from an interdisciplinary perspective to gain a deeper understanding of the environment and ecosystem. The engineers alone may not [be able to] understand and address different problems and dimensions related to the delta. Delta management is not about dealing with large infrastructure but is also essential to deal with the ecosystems and related services. Hence, professionals from various

Table 2 | Themes and sub-themes emerging in interviews.

	Major themes				
	Theme 1 **Potential delta professionals and stakeholders, as perceived**	**Theme 2** **Required knowledge, skills, and attitudes**	**Theme 3** **Senior and junior delta professionals**	**Theme 4** **Organisational culture and capacity**	**Theme 5** **Educational capacity**
Sub-themes	Involved in project formulation and implementation related to water resource management (e.g., infrastructure, organisational capacity, research, and training)	Expertise in a broad range of disciplines (e.g., social sciences, human behaviour, natural science, and technology)	Blending experiences of young and professionals	Need to foster quick and efficient coordination among organisations (government, NGOs, private sector)	Still predominantly theory-based water education (lack of practical orientations and creativity development)
	Multidimensional perspective (e.g., agriculture, environment, economics, sociology, gender, ecosystem, biodiversity, etc.)	Adaptation capacity of knowledge and information from the global perspective	Different types of responsibilities/ contributions	Autonomy and institutional capacities	Introduce demand-driven disciplines and courses in educational institutions
	People from different professions and backgrounds (e.g., policy planners, programmers, researchers, civil society, communities)	Reshape global information in line with the local needs and context		Formation of a powerful coordinating body	Learner/student-oriented teaching
				High-level autonomy and empowerment of such coordinating body	

disciplines and backgrounds are the key requirement' (P1). Similarly, another participant reflected '[The] Delta plan is a multidimensional approach. Although it focuses on the water sector, it touches upon social, environmental, gender, ecosystem, and biodiversity aspects. [...] individuals and/or groups having expertise in these sectors can be the potential delta professionals' (P12). Another participant added 'If you focus on the engineers only, it will be a big mistake. You will have to involve experts and people from other disciplines such as agriculture, environment, economics, sociology, and so on because water is a cross-cutting issue. It touches everything' (P2).

Another participant explained how a team comprising people from different disciplines and backgrounds would likely be able to contribute to the design and implementation of projects: 'Engineers are designing many constructions in the haor[1] region for better water management and protect rice from flash flood. But the engineers have little or no collaboration and communication with the agronomist. Recently, locals with expertise felt that there is a potential to develop a new variety of rice that can be harvested two weeks earlier than the ones

[1] A *haor* is a wetland ecosystem in the north eastern part of Bangladesh which physically is a bowl-shaped shallow depression, also known as a back-swamp.

they are cultivating now…thus, the agronomists can contribute. But, we did not think like this two decades ago' (P1). Another participant believed that delta water management is the central focus of the BDP2100. Water is the critical element for delta agriculture, environment, fisheries, forestry, and much more that is essential for the delta's livelihoods. Therefore, the local stakeholders with deep local familiarity with these sectors are potential 'delta experts'. However, the interviewees specifically identified four groups of people as the probable delta professionals or experts. The first group comprises professionals who understand water science, notably biology, chemistry, physics, applied sciences, and water engineering. The second group consists of those who are involved in water-related policy formulation. The third group will likely implement the delta plan at the field and community levels. And the fourth group consists of the communities themselves, who are perceived to have gained extensive knowledge and experience in dealing with water-related issues throughout their lives. A participant stated 'For example, I am a water engineering expert. It does not necessarily mean that I have a better understanding about water issues than a farmer because he as farmer has a life-long experience on this issue' (P5). The importance of involving the community was reflected by other participants: 'Ultimately, the Delta Plan will be implemented at the community level. Therefore, communities need to be involved in this process. If the community does not understand what is being implemented, they will probably not respond better and more effectively. […] Why are such interventions there? Why not other areas? These questions need to be understood by the communities. […] they should be involved in the planning process' (P8). This view was reiterated by another participant who suggested two groups of potential delta professionals/experts exist: a first group are the citizens (communities) living in this delta, who may have no or little formal education, even may be illiterate; the second group consists of professionals with diverse educational backgrounds and disciplines. The participant stated 'They (community members) do not read Bangla, but they live in this delta. Nobody else lives in this delta – no engineers, no […] professors, no delta planners. […] the communities that do live in this delta are the "delta professionals". They have a different kind of knowledge – unwritten, not structured. But they are the majority and the important professionals of this delta if you want to understand the delta' (P3). Furthermore, a participant clarified how and whether community members can contribute to delta management initiatives: 'The Water Board has installed about 2,700 sluice gates across the country. Of them, nearly 2,300 do not work. Why did this happen? […] because they [the Water Board] did not understand what people wanted. What are the people's needs? What do people think? How to include them? This is a great failure. […] If you do it without going to local communities and understanding their expertise, it will create trouble for the future' (P3). The participant further noted that delta plans in other countries have included the communities, which contributed much to that process: 'When the original delta experience was developed in the Netherlands 200/300 hundred years ago, these types of people [community members] have played roles because the level of professionals from the universities with experimental scientific knowledge was not yet available. […] That may be one of the reasons why they have succeeded' (P3). Another participant shared this: 'It has now been established that an individual with the life experience is the best expert. Regarding the delta-related issues, we need to consider the community as they are experiencing so many things throughout their lives. […] communities can be a potential "delta professional"' (P4). However, one participant suggested that the 'potential delta professional' should be considered from a broader perspective: media personalities, local-level politicians (e.g., members of local government) and human rights activists may be considered delta professionals. They may contribute to policy formulation and advocacy and help reflect the community's voice in many instances: 'A wide category of people such as researchers, individual water resource specialists, government administrators, policy planners, gender specialists, human rights activists, and media personalities, might all well be considered as "delta professionals"' (P9).

On the other hand, seven out of 16 participants reflected that the Delta Plan is water-centered, and that, therefore, the primary focus should be on those who do possess experience and knowledge of the water sector proper. The 'potential delta professionals' are those who are involved in the water-related sectors: '[...] The Delta plan is considered a water-centered plan; therefore, people involved in water resource management organisations such as the Bangladesh Water Development Board and the Water Resources Planning Organisation of Bangladesh should be in the forefront as delta experts' (P16).

(2) *Required knowledge, skills, and attitudes*

More than two-thirds of the participants stated that the delta professionals' knowledge, skills, and attitudes are crucial for developing, implementing, and monitoring the Delta Plan. Professionals currently involved in this process lack this knowledge and these skills because this approach is new to Bangladesh. Many aspects of delta management (e.g., sedimentology, environmental issues, river morphology, and surface and groundwater modelling) remain underexplored. Furthermore, the interviewee thought that the professionals would likely be exposed to many new challenges during the implementation in the coming years. One participant stated 'The Delta Plan has a 100-year time horizon. Many issues relating to this plan remain unexplored or underexplored. But, we will likely be exposed to many new issues over time. Exposure to new issues will develop new knowledge. For example, we built many polders in the southern part of Bangladesh. At that time, we did not have adequate knowledge about environmental issues. Now, we are experiencing continuous water logging in that region. Thus, exposure to new issues/aspects ultimately helps spur better understanding' (P1).

Most interviewees believed that delta professionals need to explore global experiences and expertise, and customise them in accordance with the local context. They further stated that collaboration and learning opportunities between the different disciplines, institutions, and groups would dramatically improve the level of performance, enhance the overall BDP2100 experience, and gain the ability to adapt worthwhile global knowledge to the Bangladeshi context. Another participant stated 'Adapting knowledge from global experience is important. There are so many deltas plans globally. But we need to know how to adapt this knowledge and reshape it per our context. [...] simultaneously, the collaboration between various disciplines and institutions should be emphasised in this process' (P6). Furthermore, the participant revealed that understanding and knowledge about the local context (e.g., on the rivers) is essential in delta management. Or, 'The nature (physical world) of rivers in Europe and Bangladesh is different. Rivers in Europe are controlled, but in Bangladesh they remain uncontrolled. Therefore, we need a delta professional who has such contextual knowledge' (P6).

The participants opined that project planning is an essential challenge at all levels. Project planning involves professionals from various disciplines such as agriculture, environmental science, and economics. Therefore, these professionals should have adequate training and orientation about project-related problems in different planning and implementation stages. They further mentioned that every step (i.e., designing, reviewing/approving, implementing, and monitoring) is equally essential. Therefore, delta professionals should gain knowledge relating to each aspect. They also emphasised that knowledge on more effective planning is the first priority. The delta professionals should understand possible future scenarios in delta-related issues. One participant echoed 'We [should] have a mature group of people able to devise a constructive strategy for Bangladesh to face the future' (P1).

However, some interviewees thought that as activities in a delta plan involve multiple issues and dimensions, delta professionals should have a more holistic knowledge about the delta and the people living in it. Delta professionals need to gain an understanding of some core issues that impact the livelihood of the delta's inhabitants, such as agriculture, environment, and disaster management. Still, one participant reflected that knowledge on optimum use of surface water, renewable energy, and silt management, are the critical aspects; thus, the development of evidence-based knowledge needs to focus on these aspects: 'Knowledge of the optimum use of renewable energy (solar energy/wind energy) and surface water is important because we are not likely to extract groundwater using fossil fuel energy for much longer. So, we need to develop knowledge about renewable energy. [...] knowledge about silt management should also be focused on in the Delta Plan' (P12). Two participants found that knowledge on sedimentology is vital in this process since it has a greater impact on *char* (island) formation: 'We need to understand sedimentology, its process, and implications. I feel there is still a deficiency in this issue' (P8).

Five out of 16 participants revealed that the communities in this delta are seriously affected by the hydro-meteorological and geological factors that create vulnerabilities to natural disasters. Considering this perspective, the delta professional should have solid knowledge on whether and how natural disasters affect the livelihood of the delta communities. Furthermore, the participant clarified that as the Delta Plan covers various aspects of delta formation, we need to understand this process. Nearly one-third of participants stated that delta professionals need to know the environmental, economic, and societal dimensions, including impacts on and interactions with local livelihoods. One participant stated 'There will be some [negative] impacts if we implement any projects (such as the Delta Plan). So, we need to know how to minimise the negative impacts on the environment which an environmentalist can assess. [...] We need to know the cost-benefit analysis that an economist can calculate. [...] We need to know the social cost of this project that social scientists can describe. Thus, we need to draw upon multiple disciplines' (P9).

Regarding the attitude of delta professionals, more than one-third of the participants reflected that some staff working with government agencies appear reluctant to consider opinions and views from other professionals: '[...] you will see that the entire plans are done by the engineers of the Water Development Board alone. In consequence, it causes many problems. [...] The problem is that our attitudes probably do not allow us to appreciate opinions from other organisations and disciplines' (P1). The participants mentioned that plans should be developed based on scientific research and a proper stakeholder consultation process. However, the participants believed that there is a scope for reviewing and updating the Delta Plan through stakeholder participation at different junctures. Such modification will likely develop relevant new understanding, knowledge, and skills.

(3) *Senior and junior delta professionals*

More than two-thirds of the interviewees thought that the teams implementing the Delta Plan should aim to make experienced (senior) and less-experienced (junior) professionals work together. The following quotes reflected this: 'There are many layers in an organisation. They are equally important. [...] I think more importantly all of these need to understand the issues they are working for' (P6). And 'Combined teams will be needed. They can be a mixture of cadres – people who have experience in policy formulation, project design (senior/experienced), and knowledge of new technology (young professionals)' (P10).

Some participants opined that every organisation should adopt the policy to continually recruit and train new employees at the professional levels so they can take the lead in the future. Continuous replenishment of the workforce is vital so that no vacuum in expertise and leadership occurs in the future. A participant stated 'I can tell you about a crisis [...]. The Department of Forestry has not included or recruited any young professionals in the last 17 years. The Department has come to an absolute crunch for lack of young professionals able to take over in five years. [...] And by that time, the world has changed. [The older generation of] Forestry staff probably knew [paper] maps but not GIS. But everything is now based on GIS-based analysis. [...] New knowledge has to be accommodated' (P3).

On the other hand, some participants opined that the older (experienced) staff should effectively formulate projects and the younger staff should contribute to the implementation process. One participant reflected 'Senior people should be engaged in planning, while young professionals can participate in implementing interventions/ projects' (P14). One participant viewed that young professionals need to be given more importance because they will serve longer: 'The experienced professionals will likely retire much sooner than a younger one. The younger will have to take the lead' (P1). However, the participant cautioned that the most critical aspect is the involvement of younger staff in the project development. Under the existing organisational culture, the senior professionals tend to be reluctant to provide adequate space for new staff to participate in project development and learn from it. Too often, young staff remain outside project formulation and implementation since the seniors are not inclined to include them. Thus, the participants emphasised that young professionals should be part of a learning process. One participant focused on the mid-level professional, stating 'The Delta Plan is a long-term plan. A person who, for example, has only five years' service left will retire soon. Then, who will take over? Often, we need to hire the same, retired person but now as a consultant. This cannot be effective. [...] We need to focus on the capacity of mid-level employees because they will provide services for a long time' (P5).

(4) *Organisational culture and capacity*

The participants felt that the organisations involved in the Delta Plan need to strengthen their capacity for implementation. The majority of the participants stated that many unexpected project-related challenges are due to emerge during the execution of the Plan. The respective organisations will have to respond to these, and thus these organisations need to strengthen their ability to absorb new information and adjust swiftly and effectively. One participant explained that 'For example, there is no climate change cell [section/division] in our organisation. But it can be very relevant in the coming years in the course of changing climatic context' (P1).

The majority of interviewees expressed concern about the organisational culture since many organisations (predominantly, government agencies) operate as complex bureaucracies. Organisations need to adopt a culture able to support faster decision-making. Communication and coordination between the different government portfolios and departments is presently considered lacking. Furthermore, organisations that have the mandate for policy formulation often lack adequate autonomy; they remain highly dependent on the ministries. As a result, staff working with these organisations may lose motivation and enthusiasm: 'If you look at the Water Resources Planning Organisation, you will probably understand the situation: it is expected to function independently and play a dynamic role in the water-related policy domain. [...] But there is a weakness in this regard' (P2).

One participant reported that the government institutions lack adequate capacity because many positions with decision-making authority under several ministries remain vacant, and the existing workforce cannot develop its skills. The participant concluded that the government organisations need to be strengthened to allow proper implementation of the Delta Plan. Three participants (out of 16), however, thought that key organisations already possess significant ability to implement the Delta Plan, and point at the BWDB which has extensive experience in water resource management. They also reported that some employees of these organisations may lack formal academic qualifications but are nonetheless technically sound thanks to long years of learning experience on the job. They also pointed out that, on the other hand, many staff have academic degrees but lack practical exposure and understanding of real-world issues.

Concerning the implementation of the Delta Plan, the participants reported a lack of effective communication among institutions and stakeholders, especially among the governmental institutions. Consequently, some issues fail to gain attention, and costs rise. One participant stated 'I notice, even in my organisation, that one division is reluctant to communicate [with] another one regarding issues in which they both have stakes. It happens because of the organisation's culture. I think intra-organisation communication needs to be enhanced for better performance and cost-effectiveness' (P1). The participant reflected that a Delta 'cell', 'commission', or 'centre' could facilitate more effective communication and coordination between the multitude of stakeholders, agencies, and organisations to take better, cost-effective initiatives. However, the participants are concerned about the positioning and the stewardship of such a new body. Some expressed concern that the most potent stakeholders and partners (predominantly, the government agencies) may be interested in leading such an entity on their own, still ignoring the views of smaller players. One participant mentioned 'We have been trying to formulate a flood inundation project in the coastal region to help forecast the potentially flooded area caused by cyclones. [...] The BWDB claimed the sole authority since they are mandated to deal with flood forecasts. [On the] other hand, the Bangladesh Meteorological Department also claimed this authority as they are mandated for cyclone forecasting. [...] Ultimately, we failed to create that project' (P1). Five (out of 16) participants claimed that if the Delta authority remains under the Prime Minister's Office (PMO), every department and agency will participate actively, and the above problems could be averted: 'I believe every department and agency will collaborate if the PMO coordinates the process' (P1). Another participant said 'It should be headed by a minister or state minister under the PMO whose key responsibility is [to] make this document live in [its] true sense. The commission should receive the adequate budget for its operation and report to the PMO' (P2).

The key focus for such a coordinating body should be its level of autonomy and empowerment: 'A potent authority will be needed to deal with all activities related to the Delta Plan because it has to collaborate with all these ministries and departments to make it more functional' (P15). One participant suggested that such bodies need to be independent from the bureaucracy: 'No bureaucratic institution will work. [...] The proposed centre or commission should truly be independent' (P7).

(5) *Educational capacity*

The participants expressed concern about the current weak institutional capacity for creating a skilled and qualified workforce to implement a Delta Plan which would likely necessitate a dedicated contingent of experienced and knowledgeable staff; conventional training of existing staff would be insufficient. Given the longer-term horizon of a Delta Plan, this needs to be addressed foremost by the higher education system. Universities and higher educational institutes should be encouraged to introduce new relevant subjects and courses aiming to create a cadre of influential 'delta professionals' with the right blend of knowledge, skills, and attitudes. One participant clarified that 'Certain aspects related to the Delta Plan should be included in the [curricula of the] universities to help develop a new generation of staff with the appropriate knowledge. Graduates of such programmes should understand and have good exposure to adaptive delta management' (P1).

Some participants were of the opinion that the education system in Bangladesh will struggle to educate 'delta professionals' since the education system has not kept pace with current developments in the field. Furthermore, the system does not sufficiently motivate students to be engaged, such as taking part in discussion during the learning process. Such lack of critical disposition hampers acquisition of new knowledge and the development of the right attitudes. A participant reflected that 'We are not updating our education system to the requirements of the individuals, the society, or the country. We are teaching using 30 years old syllabi and still think that we can produce modern, inquisitive graduates! Knowledge has moved on. [...] but we are still frozen in an older world' (P2).

Students are perceived as having settled in this lack of motivation for questioning, and they tend to not consult the original textbooks but are satisfied with the notes provided by the lecturers. Some interviewees would advise universities that they need to develop foresight on the disciplines and courses of relevance for future delta management. The universities and other education institutes should focus on the market demand and anticipate the changes. Participants also offered advice on how universities and educational institutions can improve curricula. They should seek to incorporate new topics that are emerging in the context of changing situations in society. The universities should have standard policy to continually examine what type of human resources are needed by the country in evolving contexts and how they can help develop these resources. One participant pointed out the weak capacity in educational institutions such as universities, colleges, and research organisations, to initiate programmes that are essential for implementing a Delta Plan. For example 'There are many universities in Bangladesh, but none of them offers an undergraduate programme in hydrology. Rather, they offer some fundamental courses in hydrology under civil engineering or hydraulic engineering. However, globally many universities offer such an undergraduate programme which is very important for better water resource management. Many students go to India or The Netherlands to pursue Master's degrees in hydrology. But I believe there is no more suitable country than Bangladesh for studying hydrology. It is our failure' (P2). Another participant reflected that 'Our knowledge has to be appropriate. I don't know which discipline would be best for this huge complex idea of delta plans over a long period. But geology, engineering, water resources, soil chemistry and sedimentology, all matter' (P3).

A few participants added that in Bangladesh a severe shortage exists of trained professionals in water resource management. The country, thus, has come to depend heavily on international and national consultants. The people working in this sector are too exclusively focused on the theoretical aspects instead of on the practical applications of those theories; therefore, they miss an adequate practical, real-world orientation: 'I would say that there is a huge knowledge gap among us. However, some experts do consultancy in this sector who possess a good level of expertise. [However] Truly speaking, they may lack [continued] commitment as they are not permanently holding any positions in the organisations they work for' (P2). Some participants advocated to make education more problem-oriented and geared to real-life situations: 'Our education is not practice oriented; it rather focuses on theoretical aspects. We should be focused on practice-oriented and problem-solving education. [...] We should think about the future need and act accordingly' (P7). 'Young professionals need to be engaged in a real-world situation, toward problem-oriented approaches. There should be more internship opportunities during the educational programs [so that they can experience real-world situations]' (P8). 'We lack problem-oriented education. Our education system is heavily biased towards theory exams. [...] Our education seems not application-oriented' (P5).

Nonetheless, one participant reported that the educational institutes' capacity for addressing new thematic issues related to water resource management has recently increased as many new education programmes in this area have been started: 'Some universities in Bangladesh are offering new programmes and courses such as meteorology and oceanography' (P2). However, he opined that offering new courses and programmes may not suffice. Instead, the overall quality of education has to be improved, and significant investment is needed in equipment and laboratory facilities: 'We have a sufficient number of educational departments, but the concern is whether they are producing a qualified workforce. We need to focus on that question' (P14).

DISCUSSION

This study explores the future capacity requirements for implementing BDP2100 by analysing structured interviews with selected key water professionals. The formulation of and debate on BDP2100 are relatively new endeavours for GoB as it approved the Plan in 2018 as a water management strategy for the next decades (Bangladesh Planning Commission, 2018). Thus, many implementation aspects of BDP2100 have yet to be further examined or operationalised to suit the Bangladeshi context (Hasan *et al.*, 2020). A critical data gap exists in the literature that limits comparison and transposition modality between different country contexts. We found that capacity development is not yet integrated into BDP2100 although an array of factors related to capacity deficits will significantly affect the implementation of BDP2100.

The outcomes of this study suggest that BDP2100 implementation will likely create the demand for professionals from a diversity of disciplines, notably hydrology, civil and hydraulic engineering, social sciences, economics, shipping, fisheries, environmental science, forestry, livestock husbandry, agronomy, etc. They also confirm the value of structural involvement of 'expert stakeholders' with deep familiarity with local land and water features, habits and potential coping strategies. Earlier studies also concluded that structural interaction and collaboration between actors and parties from diverse backgrounds and roles can increase the understanding in the social, ecological, and economic aspects of deltas (Pekkarinen & Harmaakorpi, 2006; Sol & Wals, 2015). Many studies reflecting different settings demonstrated that participatory multi-stakeholder approaches facilitate innovation and improve community capacity for sustainable delta management (Blokland *et al.*, 2009; Cundill, 2010; Warner, 2010; Brouwer *et al.*, 2016). Jønch-Clausen & Fugl (2001) argued that working with a broad range of stakeholders and parties, and of management levels, correlated with better water resource management. Their study highlighted the value of the integration between and across different elements of the social and natural systems. Such integration enables stakeholders and institutions to be involved in decision making in the planning and design processes. Mcdonnell (2008) observed that the inter-connectedness of professionals from different domains helps increase the deeper understanding of the social, political, and cultural dimensions of water, eventually enhancing water resource management. Our data also suggest that interviewees concur on the importance of community involvement. Historically, community involvement has not been a high priority in policy planning and implementation in Bangladeshi delta management initiatives (Kulsum *et al.*, 2020, 2021).

A relatively small proportion of study participants thought that collaborative work (and learning opportunity) between individuals, professional groups, and institutions would effectively contribute to successful implementation of the Delta Plan. Studies in other contexts noted that collaboration among different disciplines and institutions does advance new knowledge in water and environment domains which, in turn, helps implementation effectiveness (Pohl, 2005; Harris & Lyon, 2013; Holm *et al.*, 2013; Rice, 2013). Collaboration between natural, social, and technical scientists was perceived as beneficial for better water resource management, in line with other studies (Holm *et al.*, 2013). Studies conducted in different contexts also noted that interdisciplinary and interprofessional collaborative work and learning raise the potential to enhance the effectiveness in implementing complex programmes that encompass, e.g., natural resource management, land use planning, drainage management, surface and groundwater management, livelihood development, agriculture and environmental management, and disaster management (Abdeldayem *et al.*, 2005; Mollinga, 2009). Building on this experience and taking account of the outcomes of our study, it is suggested to develop an operational framework in which a broader spectrum of knowledge, skills and attitudes on water management can be mobilised and reflection and interaction on different insights and opinions can be structured and formalised.

The participants in the study concur on the necessity to compose teams that are also heterogeneous in seniority. A combination of senior (experienced) and junior (young professionals) team members is assumed to strengthen the sustainability of projects and ensure the continuity of institutional knowledge. However, literature on this is limited. From Bangladesh's perspective, such a mix of the workforce in teams is a considerable challenge as young professionals are still rarely actively involved in project formulation and execution, which also acts as a barrier to their professional development.

The study's outcomes highlight the need for a positive organisational culture that facilitates cooperation in and across the organisations, departments and stakeholder groups, and enables adaptive planning. Grigg (2008) found that institutional barriers often are a cause of poor project outcomes. This likely applies to Bangladesh where government agencies and authorities function in a complex bureaucracy, and more conducive intra- and inter-organisational cultures may be a condition for better water management and governance (Alam & Quevauviller, 2014; Alam *et al.*, 2018). Effective communication and coordination among stakeholders (notably government agencies, NGOs, and private organisations) will likely help improve overall organisational capacity, a prerequisite for the implementation of the Bangladesh Delta Plan (Kulsum *et al.*, 2020; Mutahara *et al.*, 2020). Akpabio *et al.* (2007) and Akpabio (2008) arrived at similar conclusions for the Cross Basin River, Nigeria, where insufficient staff capacity, inadequate management systems, and closed-door politics were factors negatively affecting the integrated water resource management projects.

Pathirana *et al.* (2012) showed that often water professionals lack a practical problem-orientation because the traditional pedagogical approaches in water education are teacher – rather than pupil-centered. Also, the outcomes of the current study reveal that water education in Bangladesh tends to substantially limit learners' orientation to practical situations and problems limiting their opportunities to enhance their understanding of, and capacity for, problem solving. Instead, the prevailing teaching methodology tends to prioritise concepts and theories. As a result, the water professionals emerging from such education are inclined to focus on the theoretical perspective of that particular discipline, often limiting their capacity to comprehensively analyse the issues from a practical perspective. Thus, the pedagogic methodologies would need to be adjusted to enhance problem-oriented learning heuristics.

To help initiate, oversee and implement the meaningful involvement of the interest groups and government agencies with their diverse disciplines and backgrounds, a high-level coordinating body is recommended by a majority of respondents in this study. Other studies (Loucks, 2003; Pahl-Wostl *et al.*, 2007) proposed similar recommendations, i.e., to ensure unbiased and effective communication and collaboration between multiple, and often competing, interest groups. Our study's outcomes also suggested that such a body is able to define and demarcate specific roles and functions between the stakeholders and parties such as during design, implementation, and monitoring of outcomes.

This study has limitations. The findings of this study were based on a comparatively small sample size. We were unable to include participants from outside the domain of the water and water-related sectors in Bangladesh due to logistical reasons such as the study resources and availability of participants. Conducting interviews with a more diverse group of professionals (notably policy analysts, resource economists, and water resources lawyers and environmentalists) and 'expert stakeholders' deeply familiar with deltas (notably community representatives, environmental activists, and civil society members) would add insight and depth. At the same time, we are confident that the participant selection, the interview design and analysis, and the obtained responses are representative of the opinions and perceptions in the key professional and government circles.

CONCLUSION

The implementation of the BDP2100 requires several specific measures for capacity development in a two-pronged approach. Firstly, the capacity of individual professionals – in the sense of their knowledge, skills, and attitudes that combine into an operational capability – needs to be strengthened so that this can contribute to the development of a new community of specialised 'delta professionals'. Importantly, this operational capability should encompass skills and organisational procedures for more effective communication and collaboration between the different disciplines, as well as with the other 'expert stakeholders' at local and national levels. Secondly, a change at the organisational scale is necessary driving structural inter- and intra-agency collaboration to secure collaboration in a long-term, adaptive implementation framework, and generating a greater degree of dynamism. These two prongs are correlated and will reinforce each other.

Professionals are broadly called to develop a continuous, structural learning culture that is embedded in the team compositions and operational procedures in the delta management organisations, knowledge institutions, and institutions of higher learning.

ACKNOWLEDGEMENTS

This research was funded by Nuffic: the Dutch organisation for internationalism in education, under the project DeltaCap (project code: INNOCAP/BGD/278). The funders had no role in study design, data collection and analysis, decision to publish, or preparation of the manuscript.

CONFLICT OF INTEREST

The authors declare that they have no competing interests.

DATA AVAILABILITY STATEMENT

Data cannot be made publicly available; readers should contact the corresponding author for details.

REFERENCES

Abdeldayem, S., Hoevenaars, J., Mollinga, P. P., Scheumann, W., Slootweg, R. & Van Steenbergen, F. (2005). Agricultural drainage: towards an integrated approach. *Irrigation and Drainage Systems 19*, 71–87.

Akpabio, E. M. (2008). Integrated water resources management in the Cross river basin, Nigeria: how can we reconcile institutional boundaries and interests? *International Journal of River Basin Management 6*, 267–276.

Akpabio, E. M., Watson, N. M., Ite, U. E. & Ukpong, I. E. (2007). Integrated water resources management in the Cross River Basin, Nigeria. *Water Resources Development 23*, 691–708.

Alam, M. M. & Quevauviller, P. (2014). An evaluation of Integrated Water Resources Management (IWRM) activities in Bangladesh. *Asia Pacific Journal of Energy and Environment 1*, 22–38.

Alam, G. M., Alam, K., Mushtaq, S. & Leal Filho, W. (2018). How do climate change and associated hazards impact on the resilience of riparian rural communities in Bangladesh? Policy implications for livelihood development. *Environmental Science & Policy 84*, 7–18.

Ali, A. (1999). Climate change impacts and adaptation assessment in Bangladesh. *Climate Research 12*, 109–116.

Bangladesh Planning Commission (2018). *Bangladesh Delta Plan 2100 (Abridged Version)*. General Economics Division, Bangladesh Planning Commission, Government of the People's Republic of Bangladesh. Available at: https://oldweb.lged.gov.bd/UploadedDocument/UnitPublication/1/756/BDP%202100 %20Abridged%20Version%20English.pdf (accessed 30 May 2021).

Blokland, M., Alaerts, G., Kaspersma, J. & Hare, M. (2009). *Capacity Development for Improved Water Management*. CRC Press/Taylor & Francis, London.

Braun, V. & Clarke, V. (2006). Using thematic analysis in psychology. *Qualitative Research in Psychology 3*, 77–101.

Brouwer, H., Woodhill, J., Hemmati, M., Verhoosel, K. N. & Van Vugt, S. (2016). *The MSP Guide: How to Design and Facilitate Multi-Stakeholder Partnerships*. Practical Action Publishing, Rugby.

Cundill, G. (2010). Monitoring social learning processes in adaptive comanagement: three case studies from South Africa. *Ecology and Society 15*(3), 28–34.

De Heer, J. (2020). *Strategisch Management Voor Bangladesh Delta Plan 2100*. Available at: https://www.twynstragudde.nl/inzichten/strategisch-management-voor-bangladesh-delta-plan-2100 (accessed 24 January 2022).

Dewulf, A. & Termeer, C. (2015). Governing the future? The potential of adaptive delta management to contribute to governance capabilities for dealing with the wicked problem of climate change adaptation. *Journal of Water and Climate Change 6*, 759–771.

Eckstein, D., Künzel, V. & Schäfer, L. (2021). *Global Climate Risk Index 2021. Who Suffers Most from Extreme Weather Events, 2000–2019*. Germanwatch, Bonn/Berlin.

Fereday, J. & Muir-Cochrane, E. (2006). Demonstrating rigor using thematic analysis: a hybrid approach of inductive and deductive coding and theme development. *International Journal of Qualitative Methods 5*, 80–92.

Giosan, L., Syvitski, J., Constantinescu, S. & Day, J. (2014). Climate change: protect the world's deltas. *Nature News 516*, 31.

Goosen, H., Hasan, T., Saha, S., Rezwana, N., Rahman, R., Assaduzzaman, M., Kabir, A., Dubois, G. & Van Scheltinga, C. (2018). *Nationwide Climate Vulnerability Assessment in Bangladesh. Final Draft*. Dhaka.

Grigg, N. S. (2008). Integrated water resources management: balancing views and improving practice. *Water International 33*, 279–292.

Guest, G., Bunce, A. & Johnson, L. (2006). How many interviews are enough? An experiment with data saturation and variability. *Field Methods 18*, 59–82.

Haasnoot, M., Kwakkel, J. H., Walker, W. E. & Ter Maat, J. (2013). Dynamic adaptive policy pathways: a method for crafting robust decisions for a deeply uncertain world. *Global Environmental Change 23*, 485–498.

Harris, F. & Lyon, F. (2013). Transdisciplinary environmental research: building trust across professional cultures. *Environmental Science & Policy 31*, 109–119.

Hasan, S., Evers, J. & Zwarteveen, M. (2020). The transfer of Dutch delta planning expertise to Bangladesh: a process of policy translation. *Environmental Science & Policy 104*, 161–173.

Henstra, D. (2016). The tools of climate adaptation policy: analysing instruments and instrument selection. *Climate Policy 16*, 496–521.

Holm, P., Goodsite, M. E., Cloetingh, S., Agnoletti, M., Moldan, B., Lang, D. J., Leemans, R., Moeller, J. O., Buendía, M. P. & Pohl, W. (2013). Collaboration between the natural, social and human sciences in global change research. *Environmental Science & Policy 28*, 25–35.

Huq, S. & Ayer, J. (2008). *Climate Change Impacts and Responses in Bangladesh; A Report of European Parliament. IP/A/CLIM/IC/2007-106.*

Jeuken, A., Haasnoot, M., Reeder, T. & Ward, P. (2014). Lessons learnt from adaptation planning in four deltas and coastal cities. *Journal of Water and Climate Change 6*, 711–728.

Jønch-Clausen, T. & Fugl, J. (2001). Firming up the conceptual basis of integrated water resources management. *International Journal of Water Resources Development 17*, 501–510.

Karim, M. F. & Mimura, N. (2008). Impacts of climate change and sea-level rise on cyclonic storm surge floods in Bangladesh. *Global Environmental Change 18*, 490–500.

Khan, M. R., Huq, S., Risha, A. N. & Alam, S. S. (2021). High-density population and displacement in Bangladesh. *Science 372*, 1290–1293.

Kulsum, U., Timmermans, J., Khan, M. S. A. & Thissen, W. (2020). A conceptual model-based approach to explore community livelihood adaptation under uncertainty for adaptive delta management. *International Journal of Sustainable Development & World Ecology 27*, 583–595.

Kulsum, U., Timmermans, J., Haasnoot, M., Khan, M. S. A. & Thissen, W. (2021). Why uncertainty in community livelihood adaptation is important for adaptive delta management: a case study in polders of Southwest Bangladesh. *Environmental Science & Policy 119*, 54–65.

Lempert, R. J., Popper, S. W. & Bankes, S. C. (2003). *Shaping the Next One Hundred Years: New Methods for Quantitative, Long-Term Policy Analysis.* RAND Corp., Santa Monica, CA.

Loucks, D. P. (2003). Managing America's rivers: Who's doing it?. *International Journal of River Basin Management 1*, 21–31. DOI: 10.1080/15715124.2003.9635189.

Marchand, M. & Ludwig, F. (2014). *Towards a Comprehensive Framework for Adaptive Delta Management.* Delta Alliance, Delft/Wageningen, Netherlands. Available at: http://www.delta-alliance.org/gfx_content/documents/documentation/TowardsaComprehensiveFrameworkforAdaptiveDeltaManagement.pdf.

Mcdonnell, R. A. (2008). Challenges for integrated water resources management: how do we provide the knowledge to support truly integrated thinking? *International Journal of Water Resources Development 24*, 131–143.

Mintzer, I. (1988). Living in a warmer world: challenges for policy analysis and management. *Journal of Policy Analysis and Management 7*, 445–459.

Mollinga, P. P. (2009). Towards the transdisciplinary engineer: incorporating ecology, equity and democracy concerns into water professionals' attitudes, skills and knowledge. *Irrigation and Drainage 58*, S195–S204.

Mutahara, M., Warner, J. & Khan, M. (2020). Multi-stakeholder participation for sustainable delta management: a challenge of the socio-technical transformation in the management practices in Bangladesh. *International Journal of Sustainable Development & World Ecology 27*, 611–624.

Nicholls, R., Hutton, C., Lázár, A., Allan, A., Adger, W., Adams, H., Wolf, J., Rahman, M. & Salehin, M. (2016). Integrated assessment of social and environmental sustainability dynamics in the Ganges-Brahmaputra-Meghna delta, Bangladesh. *Estuarine, Coastal and Shelf Science 183*, 370–381.

Nicholls, R. J., Hutton, C. W., Adger, W. N., Hanson, S. E., Rahman, M. & Salehin, M. (2018). *Ecosystem Services for Well-Being in Deltas: Integrated Assessment for Policy Analysis.* Springer Nature, London/Berlin.

Pahl-Wostl, C. (2007). Transitions towards adaptive management of water facing climate and global change. *Water Resources Management 21*, 49–62.

Pahl-Wostl, C., Sendzimir, J., Jeffrey, P., Aerts, J., Berkamp, G. & Cross, K. (2007). Managing change toward adaptive water management through social learning. *Ecology and Society 12*(2), 30.

Pathirana, A., Koster, J., Jong, E. D. & Uhlenbrook, S. (2012). On teaching styles of water educators and the impact of didactic training. *Hydrology and Earth System Sciences 16*, 3677–3688.

Pekkarinen, S. & Harmaakorpi, V. (2006). Building regional innovation networks: the definition of an age business core process in a regional innovation system. *Regional Studies 40*, 401–413.

Pohl, C. (2005). Transdisciplinary collaboration in environmental research. *Futures 37*, 1159–1178.

Radhakrishnan, M., Pathirana, A., Ashley, R. M., Gersonius, B. & Zevenbergen, C. (2018). Flexible adaptation planning for water sensitive cities. *Cities 78*, 87–95.

Renaud, F. G., Syvitski, J. P., Sebesvari, Z., Werners, S. E., Kremer, H., Kuenzer, C., Ramesh, R., Jeuken, A. & Friedrich, J. (2013). Tipping from the Holocene to the Anthropocene: how threatened are major world deltas? *Current Opinion in Environmental Sustainability 5*, 644–654.

Rice, M. (2013). Spanning disciplinary, sectoral and international boundaries: a sea change towards transdisciplinary global environmental change research? *Current Opinion in Environmental Sustainability 5*, 409–419.

Shahid, S. (2011). Impact of climate change on irrigation water demand of dry season Boro rice in northwest Bangladesh. *Climatic Change 105*, 433–453.

Sol, J. & Wals, A. E. (2015). Strengthening ecological mindfulness through hybrid learning in vital coalitions. *Cultural Studies of Science Education 10*, 203–214.

Swanson, D. & Bhadwal, S. (2009). *Creating Adaptive Policies: A Guide for Policymaking in an Uncertain World.* IDRC, Toronto.

Timmermans, J., Haasnoot, M., Hermans, L., Kwakkel, J. H., Rutten, M. & Thissen, W. A. (2015). Adaptive delta management: roots and branches. In *E-proceedings of the 36th IAHR World Congress*, 28 June–3 July 2015, The Hague, the Netherlands.

Walker, W. E., Rahman, S. A. & Cave, J. (2001). Adaptive policies, policy analysis, and policy-making. *European Journal of Operational Research 128*, 282–289.

Warner, J. (2010). Integration through compartmentalization? Pitfalls of 'poldering' in Bangladesh. *Nature and Culture 5*, 65–83.

Wiens, J. A., Zedler, J. B., Resh, V. H., Collier, T. K., Brandt, S., Norgaard, R. B., Lund, J. R., Atwater, B., Canuel, E. & Fernando, H. J. (2017). Facilitating adaptive management in California's Sacramento–San Joaquin Delta. *San Francisco Estuary and Watershed Science 15*(2), DOI: 10.15447/sfews.2017v15iss2art3.

Zandvoort, M., Campos, I. S., Vizinho, A., Penha-Lopes, G., Lorencová, E. K., Van Der Brugge, R., Van Der Vlist, M. J., Van Den Brink, A. & Jeuken, A. B. M. (2017). Adaptation pathways in planning for uncertain climate change: applications in Portugal, the Czech Republic and the Netherlands. *Environmental Science & Policy 78*, 18–26.

Zegwaard, A., Zwarteveen, M., Van Halsema, G. & Petersen, A. (2019). Sameness and difference in delta planning. *Environmental Science & Policy 94*, 237–244.

Zevenbergen, C., Van Herk, S., Rijke, J., Kabat, P., Bloemen, P., Ashley, R., Speers, A., Gersonius, B. & Veerbeek, W. (2013). Taming global flood disasters. Lessons learned from Dutch experience. *Natural Hazards 65*, 1217–1225.

Zevenbergen, C., Khan, S. A., Van Alphen, J., Terwisscha Van Scheltinga, C. & Veerbeek, W. (2018). Adaptive delta management: a comparison between the Netherlands and Bangladesh Delta Program. *International Journal of River Basin Management 16*, 299–305.

First received 11 October 2021; accepted in revised form 19 April 2022. Available online 13 May 2022

© 2022 The Authors From Knowledge and Capacity Development to an Implementation Science: doi: 10.2166/wp.2022.225

Capacity development for SDG 6.5 on IWRM and transboundary cooperation: opportunities and barriers

Carla Sabbatini [a,*] and Damian Indij [b]

[a] School of Education, Universidad de San Andrés, Victoria, Argentina
[b] Cap-Net UNDP, Buenos Aires, Argentina
*Corresponding author. E-mail: csabbatini@udesa.edu.ar

CS, 0000-0001-6142-1851

ABSTRACT

Capacity development (CD) is acknowledged as a driver of effective implementation of integrated water resources management (IWRM) (SDG 6.5.1) and transboundary water management (SDG 6.5.2). This paper explores the factors that facilitate or hinder CD for these agendas, to describe the roles water CD organisations play in support of these SDGs at the country level and to suggest potential pathways to leverage their work. A survey was conducted to analyse the perspectives of 41 water CD programme managers on their operational modes, strengths, opportunities and main challenges. Selected key informants were interviewed in depth, and the findings were analysed in the light of the literature on CD. The results highlight (1) an array of good-practice activities currently in progress, which provide potential for scaling up to accelerate progress; (2) wide acceptance of the relevance of the 2030 Agenda, even though this does not always explicitly address specific SDGs and (3) the difficulties in making the 2030 Agenda relevant to local communities. Further research is recommended to understand the context- and culturally-bound specifics of CD, to produce more locally sensitive and effective impact assessment frameworks and to adjust support to leverage CD to accelerate progress on the SDGs.

Key words: Capacity development, Cultural constraints, Integrated water resources management, SDG Implementation, Transboundary water management

HIGHLIGHTS

- Perspectives of capacity developers (CDs).
- Overview of the relevant contributions of CD organisations to SDGs 6.5.1 and 6.5.2.
- Challenges in localising targeted SDGs, often addressed but not explicitly for the sake of getting support from local communities.
- Weight of cultural factors in shaping CD and its impact.
- Need for further research to better understand CD and enhance its contribution to sustainable development.

INTRODUCTION

The 'Water Challenge' is acknowledged as a cornerstone of sustainable development and is reflected in the 2030 Agenda through its Sustainable Development Goal 6 (SDG 6). To address the complexity of such a challenge, agendas are being advocated globally to scale up the implementation of, *inter alia*, target SDG 6.5, which calls for integrated water resources management (IWRM) at all levels, including transboundary cooperation for water management, where appropriate (specified through two indicators, 6.5.1 and 6.5.2).[1] GWP (2007) defines IWRM as a process that promotes the co-ordinated development and management of water, land and related resources, in order to maximise the resultant economic and social welfare in an equitable manner without compromising the sustainability of vital ecosystems. While necessary, it is an ambitious and often complex task; though countries are making progress, recent reports (UNESCO, 2017; UN Environment, 2018) also show that implementation status varies widely and stress the need to accelerate progress. In this scenario of pressing urgency, underscored by the current pandemic,[2] efforts must become more effective for managing water more sustainably, supported by capacity development (CD). While key elements such as enabling environments, institutions, management instruments and financing are monitored to steer and assess progress, the specific role of CD in achieving SDGs 6.5.1 and 6.5.2 remains less well analysed, although Ibisch *et al.* (2016), among others, have demonstrated the value of aligning IWRM development processes and CD processes.

The UN Environment report (2018) points out that implementing IWRM is very locality- and time-sensitive, and this paper will argue that this applies equally to CD. A consensus has emerged about what IWRM is or should be in general terms, but less clarity exists about how to adjust and adopt it in diverse local settings. GWP's Toolbox (2021) identifies different development pathways, recognising their potentially replicable nature; still, the factors that define why a particular pathway or model is functional in specific circumstances only are not well described. Thus, this area would benefit from further investigation. Some countries are reported to have succeeded in increasing cross-sectoral collaboration, a central feature of IWRM. Others have progressed on more specific scores, such as gender objectives. Others still have been successful in implementing very practical measures of an operational nature. The UN Environment report (2018) screens these diverse achievements but devotes little analysis to how CD has supported these processes. 'Lack of capacity' is typically diagnosed in countries that perform poorly in the introduction of IWRM, and 'capacity development' is then recommended, but overall, the mechanics, role and impact of CD are not addressed beyond the suggestion that 'learning opportunities and potential peer-to-peer capacity building' should exist. Thus, although many policy statements and studies call for CD (e.g. UNDP, 2009; Christoplos *et al.*, 2014; Alaerts & Kaspersma, this Volume), surprisingly little is known about how it can drive transformational change in institutions towards the advancement of sustainable water management. This paper intends to contribute to the understanding of how CD can support IWRM development by studying how organisations developing capacities define their strategies and interventions. This will be based on a survey of their perceptions of challenge and success in this matter, which steer their decisions. Ways forward will be suggested to leverage CD for sustainable water management, particularly with respect to accelerating SDG 6.5.

BACKGROUND

A core concept underpinning IWRM is that sustainable water management requires that water is approached as a hydrologic *system* physically connecting the different supplies and uses, and the land and ecosystem resources. This physical system is complex and managed by institutions, and, thus, it is essential to produce and disseminate knowledge as well as to use this knowledge to build the capacity of water managers. However, knowledge by itself does not ensure institutional capacity to act. The assumption that greater availability of knowledge will result in linearly improved management practices is unrealistic and leads to inefficient deployment of efforts. Thus, to

[1] The SDG 6 IWRM Support Programme assists governments in designing and implementing country-led responses to SDG indicator 6.5.1, the degree of implementation of IWRM. Under the guidance of the UN Environment Programme (UNEP), this programme is co-ordinated by the Global Water Partnership (GWP) in collaboration with UNEP-DHI Centre and Cap-Net UNDP.

[2] For example, https://en.unesco.org/news/covid-19-implication-water-megacities (accessed 25 September 2020).

better understand how CD can assist in strengthening IWRM, the institutional architecture of the latter must be investigated, including its governance, policymaking, management and planning.

Current conceptual and operational approaches to CD come from various fields of social sciences as well as from case studies and field practice; sometimes these contradict one another. The concept of 'capacity' has been developed in different contexts such as organisational development, institutional economics, public administration, pedagogy and sociology (e.g. Alaerts & Kaspersma, 2019). This diversity in origin and application underscores its prevalence but also suggests complexity; however, pedagogical approaches and instruments are a common feature. CD is being increasingly viewed as a complex learning process that involves not only the transfer of knowledge and capacity from one social system to another, but also deliberate efforts by a particular system (such as an organisation) to create, strengthen and maintain its overall capacity over time (Mvulirwenande *et al.*, 2014). Awareness is also growing that in order to enhance CD effectiveness, conceptual and operational frameworks need to be further elaborated, including shifting beyond the linear cause-result understanding that inputs delivered by practitioners will be straightforwardly translated into outputs measurable in terms of key performance indicators (KPIs). This goes in line with the importance of reinterpreting and/or transforming policy and the environment in which the CD is taking place. Knowledge institutions seem to be leading in research aimed at creating such new frameworks (e.g. Pascual Sanz *et al.*, 2013). As pointed out by Scriven (1991, in Maxwell, 2012), applied social research is particularly useful to utilise the *process* through which things are happening in a situation (e.g. learning) and from there help improve existing practice, rather than *ex ante* determine outcomes of the programme or practice studied.

CD is closely related to, but differs from, knowledge acquisition. Christoplos *et al.* (2014) define CD as 'consisting of a range of dimensions, from the knowledge of individuals to that of organisations and the nature of the institutional frameworks and norms in which they operate [...] CD is about change. Intended outcomes generally involve the extent to which people perform their jobs differently. In order to enable them to do so they may need greater knowledge, bureaucracies or organisations that encourage them [...].' Thus, knowledge acquisition is a necessary but not sufficient condition for CD. Measuring the impact of CD is also challenging because of the multiple dimensions – individuals, organisations and their environments (including the framework conditions, i.e. institutions and policies, values systems and cultural patterns) – in which it operates; also measuring competences through standardised tests is complicated and is the subject of cultural context (e.g. Steiner-Khamsi, 2004; Beech, 2009). Hence, CD is both an end in itself and the means to an end, and indeed, a specific procedure (GTZ, 2003). As a means and a procedure to bring forth change towards sustainability, CD needs to be tailor-made. UNDP (2009) concludes that decades of experimenting with development models has confirmed the value of local ownership and capacity, and that technical assistance, though appropriate sometimes to address short-term needs, tends to be donor-driven, while distorting national priorities. Strong capacity, locally generated, is deemed essential to the success of any development enterprise. This study, therefore, seeks to reflect cultural idiosyncracies and other traits of local identity by exploring the extent to which current practitioners find that they enable or hinder their efforts to promote IWRM and SDG 6.5.

To sum up, 'capacity development' is a rather elusive concept built from the contributions of different fields, and it comprises an array of very diverse strategies with varying purposes; in this sense, this paper intends to illustrate such diversity by collecting examples from water CD practitioners from different organisational, cultural and geographical contexts across the globe. It does not seek to thoroughly map that diversity but rather to provide representative examples reflecting its complexity. As part of this, we shall review context- and culture-specific challenges as an aspect that needs special attention, and we shall also suggest that interdisciplinary approaches to these challenges may be suitable in attempting to address them.

METHODOLOGY

A qualitative survey was conducted with the aim of capturing the perspectives and experiences of practitioners tasked to apply CD for the purpose of developing IWRM approaches. The survey aims to better understand the complexity of CD for IWRM, how culture- and locality-specific challenges arise, and to what extent pedagogical methods prove instrumental. The survey was distributed through the networks of Cap-Net (a globally operating network under the UNDP to develop capacity in the water sector) and the Global Water Partnership (GWP) (a global network with over 3,000 partner organisations to provide knowledge and build capacity on water management at all levels, based in Stockholm). The combined networks are representative of government, non-government and multilateral agencies across the globe that have a relation with water. Of 47 organisations that were approached, 41 responses were received, though not all respondents reflected on all questions; the respondents represent key CD organisations in every major region in the world. The survey comprised 24 questions, of which 11 were open for respondents to develop their ideas, grouped into three parts: (1) personal/organisational information; (2) the CD activities carried out by the respondents' organisation and (3) the demand the respondent's organisation caters for and the partnerships it is involved in. The responses are not a representative sample of the universe of water CD organisations in a statistical sense; forming such a sample would be problematic because of the wide variation in nature, scale and scope of these organisations, in the comparatively small size of this universe. However, they are considered representative of important currents and field experiences as they originate from professional agencies with track records in water CD that are generally highly valued by their peers. Thus, they can be assumed to provide insightful input in a structured way that has thus far not been available. Further to the surveys, nine interviews were conducted in which selected respondents were requested to further elaborate on the opinions they had expressed in the open questions in the survey; this selection sought geographic and organisational representativity. However, gender balance could not be achieved as women were too few among survey respondents. The results presented refer to the survey responses, with the information and qualifications drawn from the interviews used to help interpretation.

RESULTS AND DISCUSSION

CD organisations and activities

The respondents represent 41 organisations[3] that operate at scales ranging from local to global and reside in 28 countries in every major region in the world. They stated to be partners of the Cap-Net network (41%), the GWP network (40%) or of more networks (19%), with 13% not responding. The degree of completeness and coherence of survey responses was assessed to ensure only reliable material was used; this was also considered an indicator of commitment to this study. Most survey respondents (48%) have worked for their current organisations for over 10 years, which suggests that an adequate experiential basis is reflected in responses. The respondents'

[3] ACUMAR, Cátedra UNESCO Agua y Educación para el Desarrollo Sostenible, FICH - UNL (Argentina); Cultura Ambiental, Global Water Partnership (Uruguay); Fundación Futuro Latinoamericano (Ecuador); Universidad Andina del Cusco (Peru); Universidad Externado de Colombia (Colombia); Universidad Autónoma Del Estado De México, Instituto Interamericano de Tecnología y Ciencias del Agua (Mexico); REDICA (Costa Rica); Caribbean WaterNet/CapNet UNDP/UWI (Caribbean); Country Water Partnership (Armenia); National Water Partnership (Georgia); Global Water Partnership (Ukraine); GEF IW:LEARN (France); IHE-Delft (The Netherlands); GWPO, SIWI (Sweden); CSD Center, Regional Water Partnership Central Asia, Caucasus and Mongolia (Kazakhstan); Global Water Partnership Central and Eastern Europe (Slovakia); AGW-Net (Senegal); CB-HYDRONET, CRREBaC (DR Congo); WaterCap (Kenya); Dabane Trust, WaterNet, Zimbabwe Open University (Zimbabwe); Development Workshop (Angola); Polytechnic University of Malawi (Malawi); GWP-CAf (Cameroon); Bangladesh Water Partnership, BRAC (Bangladesh); Jalsrot Vikas Sanstha (Nepal); Global Water Partnership South Asia, University of Peradeniya (Sri Lanka); CKNet (Indonesia); MyCDNet (Malaysia).

distribution by organisational role is as follows (respondents can hold more than one role): leader (senior manager, advisory board member, chair holder, CEO): 19%; coordinator (middle/senior manager, network managers/coordinator, programme leader): 51%; project manager (middle- and/or lower-level manager, administrator, executive in charge of specific projects): 21%; specialist (advisor on specific issues, technical expert): 21%; researcher: 26%; teacher (university or other lecturer, workshop facilitator): 33%.

The gender balance favours men, with about 30% female. Although caution is warranted about representativity of this group for water organisations working on SDG 6.5, this result is noteworthy. In making SDG 6 and IWRM policies, the pivotal role of women in water management is emphasised, and organisations belonging to the Cap-Net and GWP networks are specifically expected to ensure this. Although the study intended to add a comparison of professional perspectives and opinions of females and males, this ambition was curtailed by the outbreak of the pandemic; however, this subject deserves further attention.

The types of organisations represented in the survey are as follows (out of 45 responses): university/research institution: 33.3%; UN agency/international programme: 3.9%; government agency/ministry: 2%; river basin organisation: 3.9%; civil society organisation/NGO: 43.1%; other (networks of knowledge institutions, regional intergovernmental organisations): 13.7%; private-sector/consulting companies: 0%; utilities: 0%. Although IWRM and transboundary cooperation typically do not, or only rarely, involve utilities or private-sector partners, the absence of these categories may also be due to bias in the composition of the networks or the survey distribution. The target groups of the activities are as follows (as each organisation may have several target groups, respondents can identify more than one category, totalling 224): water professionals: 89%; policy-makers/government officials: 76%; community members: 62%; lecturers/researchers: 78%; school principals/teachers: 24%; higher-education students: 70%; school students: 32%; private sector: 57%; UN agencies/international programmes: 35%; NGOs: 65%; other: 16%. Others include river basin organisations, community action groups, worker cooperatives and the civil society at large. Although the salient target group is water professionals, the organisations also offer CD to a pertinent diversity of target groups. This is consistent with the nature of IWRM and transboundary cooperations that call for multi-stakeholder involvement, as well as with the recommendation that for effective CD a multi-level approach is advisable (e.g. Ibisch *et al.*, 2016). Ibisch *et al.* also stress the need for continuous and long-term engagement, particularly in processes of organisational and institutional change where no single set of guidelines or practices can be defined upfront.

The types of CD activities deployed were as follows (out of 37 responses): face-to-face short courses and workshops: 22%; online teaching: 12%; university education (undergraduate, masters, PhD): 12%; schooling: 5%; teaching materials development: 16%; awareness raising: 16%; research: 15%; other: 2%. The other activities concerned, i.e. water-related data management and dissemination, institutional knowledge management, support to local governments and basin projects, and policy briefings.

IWRM and transboundary cooperation in CD

Most respondents confirmed that their organisations' activities focus on IWRM and transboundary cooperation but that these are usually not explicitly framed as SDG 6.5. The way the SDG is addressed is reported to take several forms and rationales (Figure 1). In interviews, the great value of having a global agenda was unanimously recognised, yet it was acknowledged that 'there is less adherence at local levels' as SDGs are often 'perceived as a far-away discourse that interferes with local efforts to work on concrete solutions'. Also, some respondents report that 'national government capacity is supported to track and report on performance indicators for water resources management', 'focus on indicator SDG 6.5.1 (for IWRM) is at country level only', 'most training focuses on transboundary cooperation as a theme, not just in relation to SDG 6.5', 'we deliver activities relevant to SDG

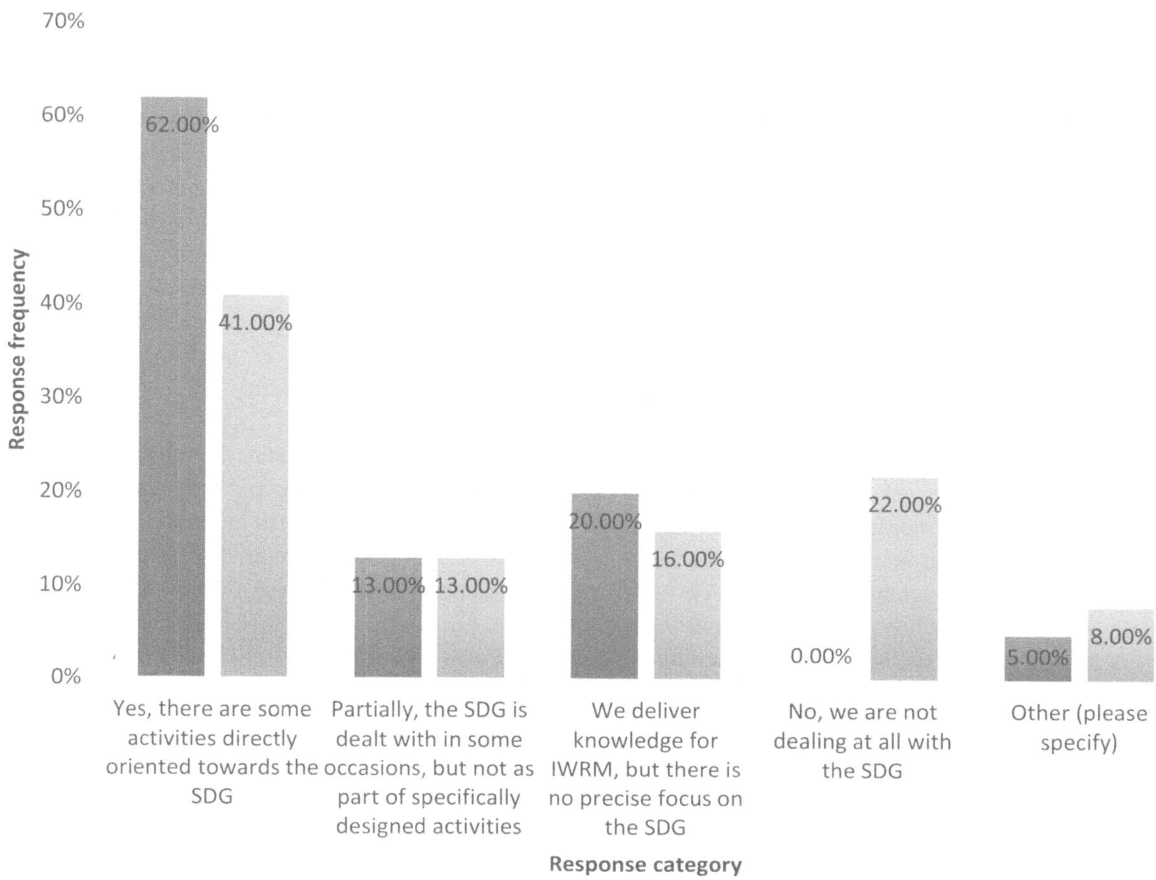

Fig. 1 | Explicit focus of respondents' CD activities to IWRM (SDG 6.5.1) and transboundary cooperation (SDG 6.5.2) (37 respondents) (dark shade: IWRM; light shade: transboundary cooperation).

6.5 but not for transboundary resources or cooperation agreements' and 'activities are joint with UNESCO IHP'. The CD activities are more aligned with IWRM (SDG 6.5.1) than with transboundary cooperation (SDG 6.5.2). Also, 22% of the respondents acknowledged no link with transboundary cooperation, while no such responses were made regarding IWRM. These results may be the consequence of the fact that IWRM is a task that pertains to all river basins in a nation, whereas transboundary cooperation concerns more particular, limited activities. However, it may be helpful to look further into these differences, as UNESCO (2017) points out the importance of SDG indicator 6.5.2.

Interviewees suggested locally specific reasons for this apparent reluctance to explicitly address SDG 6.5.2: transboundary cooperation is less generic than IWRM, fewer resources are deployed for transboundary cooperation; the lack of openness of decision-makers to acknowledge the value of CD hinders developing strategies towards attaining SDGs in general; transboundary cooperation poses more institutional challenges than IWRM; as transboundary cooperation usually involves conflicting interests from two or more jurisdictions or countries, political factors may play a stronger role; and IWRM may be perceived as more urgent/important than transboundary cooperation, and/or the latter may be regarded as a subsidiary to IWRM.

The types of CD activities used are found to be similar for both indicators. Respondents described activities that can be grouped under the following six broad categories: (1) teaching (courses, seminars, training of trainers, workshops and similar face-to-face and/or virtual activities); (2) development of training materials (designing and disseminating training materials for various audiences); (3) research; (4) fieldwork (on-site water management activities involving local communities); (5) networking (forging partnerships, initiating and/or participating in collaborative initiatives) and (6) administrative and policy support (involvement in government planning/decision-making processes, strengthening governance and institutions). Figure 2 describes their relative importance; they may be mutually not exclusive (e.g. research and fieldwork with local communities, alongside teaching and advising in policymaking). The results illustrate the diverse nature of these lines of action. CD seems generally understood as sharing knowledge in the form of teaching (face-to-face courses, lectures, seminars, workshops and virtual courses) and networking (multi-stakeholder partnerships often in platforms for sharing resources, training materials, reports, etc.), although research and policy support seem more pronounced in the transboundary cooperation case.

Although the 'tool kit' of CD appears diverse, less use was made of 'convening activities' aimed for instance at reconciling trade-offs between different competitive water uses or fostering the attitudes and skills for behavioural change to manage water more effectively. Similarly, given the importance of research as input for sustainability-oriented decision-making, more effort could be made to integrate teaching with studies. Recognising that this sample may not be representative enough to cover all ongoing practices, a risk appears that the application of CD is framed too narrowly as forms of formal instruction. Thus, other activity types should be explored (for which dedicated skills must be acquired), such as the convening of locally relevant stakeholders. In any case, CD could be applied more purposely for accurately assessing local needs (e.g. UNDP, 1997, 1998a, 1998b) and design CD activity types optimised for local use. The respondents also qualified the intensity and nature

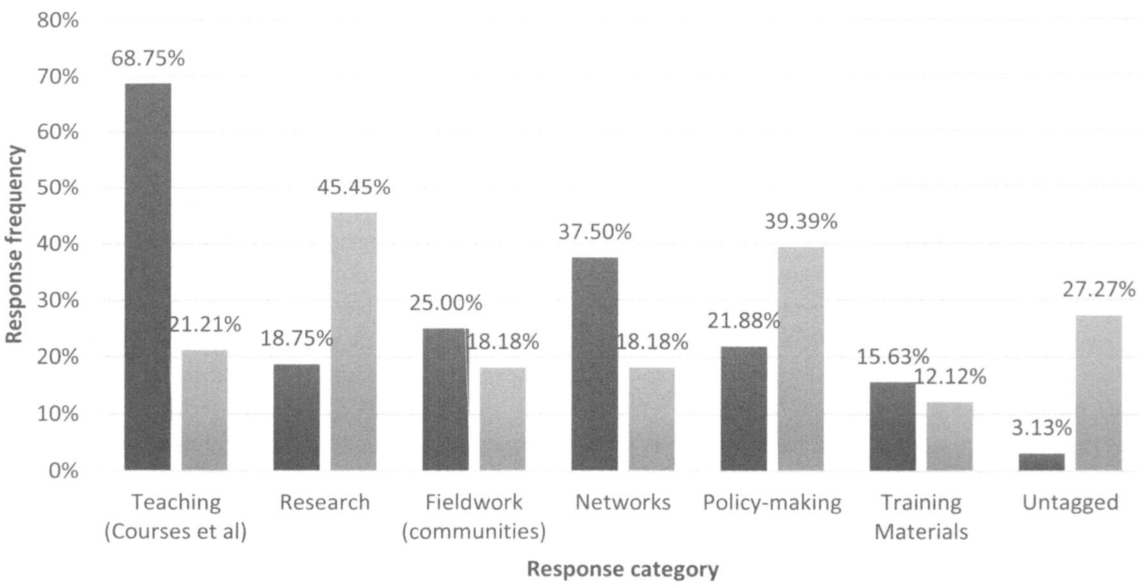

Fig. 2 | CD activities by type, in support of IWRM and transboundary cooperation (from 33 responses) (dark shade: IWRM; light shade: transboundary cooperation).

of their engagement or partnership with government water agencies (36 responses): as fluent and active (36%); with consultations and occasional joint activities (39%); with only formal meetings (3%); with little or no interaction (0%); and depending on the agency (11%). Other responses (11%) were: it is taking more time to build a relationship, or it depends on the country in which we are operating. Thus, clearly CD activities are being valued, and significant investment and long-term engagement are essential to achieve some impact.

Achievements and CD enabling factors

Over past decades, steady but modest progress was made concerning the introduction of IWRM and transboundary cooperation; although much work is yet to be done to accelerate progress, significant achievements have been made. Respondents reported a wide array of substantive contributions that can be grouped in nine categories (the percentage value relates to the number of respondents highlighting this feature):

- studies completed and data collected: 53% (e.g. on baseline indicators, data sets and as responses to policy and management queries);
- teaching: 47% (as courses and others at various levels of formal instruction, workshops, training of trainers, etc.);
- development of training materials: 16% (training/teaching resources, toolkits for various audiences);
- facilitating dialogue: 44% (creating spaces for multi-stakeholder dialogue, facilitating meetings to address sensitive issues and help resolve conflicts);
- managerial support: 31% (on financial and technical support to assist networks, countries and/or local communities);
- increased visibility: 13% (greater public recognition and reaching a wider audience);
- new policies and plans: 31% (fostered or participated in new or revised plans and policies);
- project management: 19% (devised and implemented projects) and
- partnerships: 41%.

Respondents also described key enabling factors in the survey and interviews:

- The transformational impact of IWRM itself, which stresses fundamental good-governance principles such as participation, transparency, integrity and gender balance, e.g. in the statement that 'SDG 6.5 is a natural consequence of sound [...] IWRM [...]. The SDGs are transient compared to the lasting transformational power of IWRM.'
- Sound scientific knowledge behind IWRM and for effective CD strategies. Knowledge institutions like universities and research centres can be strong enablers thanks to their reputation: 'We only propose activities, but given our reputation, countries trust that we are equipped to lead them in the right direction.'
- The leveraging power of partnerships, as expressed in statements like 'We have a strong network in the global South which allows us to both bring good practices to the fore, but also implement CD tailored to countries' needs' and 'Taking part in knowledge sharing networks such as UNESCO's UNITWIN programme has enhanced our capacity to produce and disseminate reliable knowledge applied to IWRM.' An important partnership form is that of peer-to-peer collaboration (e.g. Beck, 2018) echoed in statements like 'We have achieved more engagement with CD activities that involved local community leaders in sharing good practices, and/or working together to co-create solutions to similar water problems.'
- Increasing ability to adapt IWRM to local contexts. Respondents with a strong understanding of grass-roots, community-based applications of IWRM highlighted that this is a major enabler – even if not explicitly linking this success to the targets: 'It is our deep understanding of the local situation and the dynamics of the place that enables us to move past conflict and find ways to make IWRM principles more practicable [...].'

Constraints, barriers and challenges

Respondents reported common findings:

- *Political context can interfere with CD intentions*: This very common observation was articulated, e.g. as follows: 'We support the government but we are not the government, and [...] what we provide can be affected by political changes within governments.' And: 'The national political context is what defines the pace of progress in the implementation of IWRM.' They also reported that government adhesion to respondents' work may be on average rather low but is never non-existent and, thus, can be occasionally high.
- *Underappreciated relevance*: Decision-makers, especially at the local level, are often unaware that they need CD to understand IWRM and its benefits. A particularly strong reflection was: 'Working with politicians usually entails engaging in a battle of egos. Decision-makers resent being shown that they "need" CD–nobody likes to be told they do not know enough about their job.'
- *Narrow approach*: The multi-stakeholder nature of IWRM is fundamental, but as it is often not appreciated or is deemed too complicated, CD often reaches a too limited target audience: 'A lack of deep understanding of IWRM leads to poor coordination of water managers with other areas (e.g. in energy, agriculture or industry) which often leads to conflicting agendas.'
- *Underappreciated cultural differences*: As pointed out above, CD organisations that work closely with local communities seem to better appreciate cultural traits, and they turn cultural sensitivity to their advantage. On the other hand, CD organisations that operate more globally seem to find intercultural issues challenging: 'What in one cultural context may be understood as a well-intended gesture to suggest gratitude [...], can be regarded in another cultural context as an act of bribery or corruption [...].' And: 'What we understand by "water scarcity" is different for different people, and as a result we may need different approaches to solving a seemingly similar problem. Yet, grasping the full implications of this takes a lot of time and effort', and: 'We have to work on a listening mode, always aware that the national partner provides the context for our ideas that were hatched back at our global desks.'

• *Not enough partnerships at the regional level*: Such partnerships and platforms are key to enhance peer learning and create synergy among countries with similar water challenges and cultural identities.

- *Limited resources*: Although the limited resources allocated to CD are a common tenet, this translates specifically as follows:
  - The need to provide short-term accountability to donors often forces them to focus on activities that enable them to 'count heads' (i.e. the number of people reached by their training activities) rather than fostering deeper, longer-term transformational strategies whose impact is, however, more complex to measure. As one respondent noted, 'If we need to design rigorous, interdisciplinary frameworks to measure the extent to which our activities foster learning, we need to divert resources in this direction. [...] donors are not willing to support this, as impacts take too long to become accountable for.'
  - *Transient audiences*: Target audiences such as staff of water administrations often quickly rotate between jobs or migrate to other settings, and resources do not allow for any follow-up; this does not allow capacity to root locally, 'and we often have to start all over again from scratch'.
  - *Insufficient or difficult-to-get data*: This hampers reliable information for building, e.g. progress indicators: 'Accessing data to inform decision-making is already difficult, but even more difficult is to gather scattered or non-existing data to establish a baseline to begin with–let alone talk about measuring progress.' While this is found generally for SDG 6.5 (e.g. UN Environment, 2018), regional reports on river basin management

also show that this is particularly relevant for transboundary cooperation, where greater and early involvement of local stakeholders is instrumental in the processes of data collection and their inclusion and validation in national reports.[4]

Looking ahead: respondents' perceptions about what works and new opportunities

A set of open-answer questions in the survey, reviewed separately with each interviewee, sought to document perceptions on which activities currently in place seem to be more effective to promote SDG 6.5:

- Sustain support to CD that has proven the most effective, to achieve a significant contribution at the country level, e.g. by:
  - 'Working directly with mid-level civil servants or technical staff who are unlikely to rotate in political shifts. Also working not just with "water" ministries, but with other ministries that strongly relate to water';
  - 'Supporting strategies for adaptation to impacts of environmental changes';
  - 'Integrating IWRM in educational systems as cross-cutting at all levels of instruction';
  - 'Working with indicators and methodologies not only with the public sector but with civil society (…), using participatory approaches involving all stakeholders' and
  - 'Developing inter-jurisdictional participatory platforms for relevant issues'.
- Strengthen collaboration and global networking with other organisations. This would help scale up CD initiatives to achieve SDG 6.5 goals: 'A shared and consolidated roadmap between different CD organisations harmonising areas of focus, target audiences, etc. […]. Current efforts are dispersed and often ineffective.'
- National adoption of commitments to SDGs opens the possibility to raise awareness on the need for devising CD strategies across several sectors. Respondents pointed out that their countries are 'now engaging more in devising and implementing basin-based IWRM strategic plans applying scientific and technological advances to implement SDGs'. Similarly, local (municipal) levels of management and governance are also increasingly adopting SDGs in their agendas, which is also an opportunity for carrying out CD.

Respondents' perceptions were sought about existing capacity at the country level to implement SDG 6.5, using their professional assessment, as a function of: (1) enabling environments (law, policy, plans); (2) institutions and participation (organisational arrangements, cross-sectoral coordination, river basin organisations, public and business participation, gender objectives); (3) management instruments (water-use management, pollution control, ecosystems management, monitoring, data and information sharing, others) and (4) financial management (budgets for investment and for recurrent expenses). Table 1 shows that capacity developers rate existing capacity as medium to low in nearly all domains for both targets, with occasionally a high or very high rating. No substantive difference is reported between the four criteria, although financial management capacity appears as the most serious deficiency. These findings demonstrate, on the one hand, that more resources need to be allocated towards CD initiatives, but, on the other hand, that progress is being made on several counts, which makes a case for a greater use of peer learning.

---

[4] See, for example, Africa https://www.unece.org/fileadmin/DAM/env/documents/2018/WAT/07Jul_4-6_Douala/TDR_Atelier.pdf https://www.unece.org/index.php?id=52765 (accessed 25 September 2020), and for South America cases https://www.gwp.org/globalassets/global/gwp-sam_files/publicaciones/reporte-sdg-652-nov18/report-workshop_652-montevideo-eng-final.pdf (accessed 25 September 2020).

Table 1 | Respondents' assessment about existing capacity at the country level for SDG 6.5 for IWRM and transboundary cooperation.

	Very high capacity		High capacity		Medium capacity		Low capacity	
	IWRM	Transboundary cooperation	IWRM	Transboundary cooperation	IWRM	Transboundary cooperation	IWRM	Transboundary cooperation
1. Enabling environment	12	10	33	17	45	48	9	24
2. Institutions and participation	9	10	24	21	54	31	12	38
3. Management instruments	12	10	9	17	66	41	12	31
4. Financial management	3	4	6	11	28	25	62	61

Values are in percentages; 34 responses on IWRM, 31 on transboundary cooperation.

CONCLUSIONS

A wide range of relevant CD strategies and instruments are reported being in use, stemming from an array of organisational partnerships that each reaches out to a different target audience. This diversity is in itself a positive indicator of the commitment to global agendas like SDG 6.5 and suggests that instruments are available to suit different contexts and purposes. However, more study is warranted to further finetune and enhance the effectiveness of these CD instruments and approaches.

While the relevance of the UN's 2030 Agenda for Sustainable Development is widely acknowledged, the survey demonstrates that this is often not (yet) reflected in CD activities that are able to explicitly address specific SDGs. A general challenge is the difficulty in translating global agendas to local contexts and local communities. Better explicit alignment of CD is reported with regard to IWRM (SDG 6.5.1) than to transboundary cooperation (SDG 6.5.2), possibly because the former entails a larger and more operational agenda than the latter. The survey also suggests that the achievement of SDGs may be enhanced by a deeper understanding of the context-specific and culture-related barriers. Respondents stated that in order to reach out more effectively to local communities, at times they chose not to explicitly align their activities with the general interpretation of SDGs but recast these according to what is locally relevant. The local translation (reterritorialising) of global discourses such as the SDG agenda is actually sometimes met with resistance because of political reasons or simply because of the difficulties entailed in translating those agendas in terms of the communities' immediate needs. Water governance, management and planning are locally among the most sensitive and best-recognised aspects of SDG 6.5. Respondents also highlighted the challenge of engaging with decision-makers and administration staff who rotate between their duties and not always sustain their predecessors' achievements.

However, common 'good practices' have been identified that can be scaled up, notably multi-stakeholder participation and strong partnerships; the combination of a bottom-up community-driven groundwork with top-down governance strengthening; and the understanding of how IWRM approaches in global agendas can also foster a sense of ownership at the local level. These observations concur with recent pedagogical insights that 'educational' experiences and CD should be learner-centred, creating opportunities for peer learning as a pathway for transformational learning.

Finally, more study is required on measuring the impact of CD, distinguishing between shorter- and longer-term effects and applying appropriate frameworks to assess impact at various scales. These distinctions are conceptual and have methodological implications, for which we suggest to seek greater synergy with educational research.

FUNDING

This paper was funded as part of Cap-Net's 2020 work plan. Cap-Net serves as the global network for CD in sustainable water management. It is a UNDP delivery mechanism within its Water and Ocean Governance Programme.

DATA AVAILABILITY STATEMENT

All relevant data are available from https://drive.google.com/drive/folders/1odUrhb2J0sWAYjz76IPoRCca22uNyNGw.

REFERENCES

Alaerts, G. & Kaspersma, J. (2019). Progress and challenges in knowledge and capacity development. In *Capacity Development for Improved Water Management*. Blokland, M. W., Alaerts, G. J., Kaspersma, J. M. & Hare, M. (eds). Taylor and Francis, London, First published in 2009.

Beck, A. (2018). Water operator partnerships: peer learning and the politics of solidarity in water and sanitation service provision. *WIREs Water 6*(1). https://doi.org/10.1002/wat2.1324.

Beech, J. (2009). Policy spaces, mobile discourses, and the definition of educated identities. *Comparative Education 45*(3), 347–364.

Christoplos, I., Engstrand, K. & Hedqvist, A. L. (2014). *Capacity Development Literature Review*. UTV, SIDA, Stockholm, Working Paper 2014:1. Available at: https://www.sida.se/contentassets/e152ed3b81ab4b9ebaf51362cc2721ea/capacity-development-literature-review_3761.pdf (accessed 2 May 2020).

GTZ (2003). *Capacity Development for Sustainable Development – A Core Task of the Deutsche Gesellschaft für Technische Zusammenarbeit (GTZ)*. Deutsche Gesellschaft für Internationale Zusammenarbeit, Eschborn, Policy Paper 1.

GWP (2007). *Roadmapping for Advancing Integrated Water Resources Management Processes*. Global Water Partnership and UN Water, Copenhagen. Available at: www.unwater.org/publications/un-water-global-water-partnership-roadmapping-advancing-integrated-water-resource-management-iwrm-processes/ (accessed 6 December 2021).

GWP. (2021). *IWRM Toolbox*. Global Water Partnership, Stockholm. www.gwp.org/en/learn/iwrm-toolbox/About_IWRM_ToolBox/ (accessed 15 May 2021).

Ibisch, R. B., Leidel, M., Niemann, S., Hornidge, A. K. & Goedert, R. (2016). Capacity development for integrated water resources management: lessons learned from applied research projects. In: *Integrated Water Resources Management: Concept, Research and Implementation*. Borchardt, D., Bogardi, J. & Ibisch, R. (eds). Springer, Cham.

Maxwell, J. A. (2012). *Qualitative Research Design: An Interactive Approach*. SAGE Publications, Thousand Oaks, CA.

Mvulirwenande, S., Wehn, U. & Alaerts, G. (2014). Evaluating capacity development in the water sector: challenges and progress. In *Working Paper Presented at 37th WEDC International Conference, Hanoi, Vietnam, 2014: 'Sustainable Water and Sanitation Services for all in A Fast Changing World'*.

Pascual Sanz, M., Veenstra, S., Wehn de Montalvo, U., van Tulder, R. & Alaerts, G. (2013). What counts as 'results' in capacity development partnerships between water operators? A multi-path approach toward accountability, adaptation and learning. *Water Policy 15*(S2), 242–266. https://doi.org/10.2166/wp.2013.022.

Steiner-Khamsi, G. (2004). *Global Politics of Educational Borrowing and Lending*. Teachers College Press, New York, NY.

UNDP (1997). *Capacity Development Assessment Guidelines 2. Draft Capacity Assessment Guidelines – The Programme Approach: Assessment and Methodologies*. United Nations Development Programme, New York, NY.

UNDP (1998a). *Water: Capacity Building for Sustainable Development*. United Nations Development Programme, New York, NY, in collaboration with FAO, UNESCO, WHO and WMO.

UNDP (1998b). *Capacity Assessment and Development in a Systems and Strategic Management Context*. Technical Advisory Paper No. 3. Management Development and Governance Division Bureau for Development Policy, United Nations Development Programme, New York, NY.

UNDP (2009). *Capacity Development: A UNDP Primer*. United Nations Development Programme, New York, NY. Available at: https://www.undp.org/content/dam/aplaws/publication/en/publications/capacity-development/capacity-development-a-undp-primer/CDG_PrimerReport_final_web.pdf (accessed 20 February 2020).

UN Environment (2018). *Progress on Integrated Water Resources Management. Global Baseline for SDG 6 Indicator 6.5.1: Degree of IWRM Implementation*. United Nations Environmental Programme, Nairobi. Available at: https://www.unwater.org/publications/progress-on-integrated-water-resources-management-651/ (accessed 20 February 2020).
UNESCO (2017). *Transboundary Water Cooperation's contribution to the SDGs – Key Insights from the First SDG6.5.2 Reports*. United Nations Educational, Scientific and Cultural Organization, Paris. Available at: https://en.unesco.org/sites/default/files/hlpf652flyer.pdf (accessed 20 February 2020).

First received 7 November 2020; accepted in revised form 6 January 2022. Available online 16 February 2022

© 2022 The Authors From Knowledge and Capacity Development to an Implementation Science: doi: 10.2166/wp.2022.265

Supporting evidence-based decision-making: Capacity Building through Research

John Conallin[a,b,*], Nora Van Cauwenbergh[b,c], Nicolette Duncan [a], Win Win Zin[d], Zau Lunne[e,i], Htike Htike[f], Greg Martin[g], Thom Bogaard[h] and Mário J. Franca[b,h]

[a] Charles Sturt University, 386 Elizabeth Mitcham Drive, Thurgoona, NSW 2640, Australia
[b] IHE Delft Institute for Water Education, Westvest 7, Delft 2611 AX, The Netherlands
[c] Free University Brussels (VUB), Boulevard de la Plaine 2, Ixelles 1050, Belgium
[d] Yangon Technological University, Gyogone, Near BPI Bus-stop, Insein Rd, Yangon, Myanmar (Burma)
[e] Fauna and Flora International, 34 D/9, San Yae Twin St, Yangon, Myanmar (Burma)
[f] Karlsruhe Institute of Technology, Karlsruhe 76131, Germany
[g] Myanmar Maritime University, Thilawa, Thanlyin Township, Yangon, Myanmar
[h] Environment & Livelihoods – Governance & Program Management (Freelance Consultant)
[i] Present address: University of New Brunswick, 3 Bailey Dr, Fredericton, NB E3B 5A3, Canada
*Corresponding author. E-mail: jconallin@csu.edu.au

ND, 0000-0002-5988-2795

ABSTRACT

Lack of data inhibits informed decision-making and is a critical challenge in developing countries, many of which are under-resourced in financial, technical and institutional capacity to collect and analyse the required data. This limits the countries' ability to achieve development goals and keeps them dependent on the provision of external resources. Development initiatives often treat capacity building and research as two separate tracks of development. While efforts have been made in the health sector to combine these through project-based learning, this approach is relatively unexplored in the water sector which by its inter-sectoral nature stands to benefit significantly from a more collaborative and solution-oriented development strategy. Capacity Building through Research (CBtR) facilitates data collection and analysis by local researchers, mentored by international experts, strengthening local capacity to produce credible evidence able to inform sustainability-related decision-making. Five case studies piloting CBtR are discussed here and evaluated through criteria of the Dutch Strategy Evaluation Protocol framework. CBtR is shown to be a long-term strategy that requires the strengthening of cross-disciplinary networks to enhance the capacity of water management institutions, which likely contributes to more efficient evidence collection and analysis suitable for decision-makers, leading to greater national resilience and reduced need for external support.

Key words: Capacity development, Developing countries, Research, Strategy evaluation, Sustainable development, Water management

HIGHLIGHTS

- Project-based Capacity Building through Research (CBtR) initiatives have been piloted in the water sector and assessed.
- In CBtR, both capacity building and research are integral components of project planning.
- CBtR enhances the knowledge base and evidence for sustainability-related decision-making.
- CBtR improves national resilience, while reliance on external development organisations is reduced.

INTRODUCTION

Sustainable management of finite natural resources such as water will be essential for protecting and improving the livelihoods of human populations that rely on them. Securing a safe water supply is one challenge as many people are still lacking access to a clean, safe and reliable water supply. Water insecurity is likely to increase due to both climate change and population growth (McKenzie *et al.*, 2015). Freshwater is critical to achieving many development goals, as it is a common-pool resource that connects societal needs; it is also important for health, the environment, food production, energy, industry and domestic use (Grigg, 2016).

Recognition of the cross-sectoral demands on water resources has led to the development of the Integrated Water Resource Management (IWRM) concept, which refers to the integration of technical (engineering) aspects with management (decision-making), and societal aspects such as gender and public participation, with the use and allocation of water across sectors and scales (Grigg, 2016). Biswas (2004) criticised IWRM as a 'nirvana' concept. Though it was to some extent complemented with the water security and nexus approaches (*e.g.* Benson *et al.*, 2015), around the world it has led to the recognition that decisions on water use and allocation must be linked to the decisions in the other water-dependent sectors.

Although water has such critical integrating role across multiple sectors of society, developing countries often lack the technical expertise, governance structures and financial resources to implement IWRM. This gap is underscored by the lack of reliable data and insight about the current state of water resources; inequitable access to scientific literature; shortage of competent institutional leaders caused by limited training opportunities and inadequate support for young researchers; and often a dis-enabling policy environment which undervalues the research profession resulting in poorly defined career perspectives and inadequate research infrastructure (Hyder *et al.* 2003 as cited in Lansang & Dennis, 2004). Moreover, until recently the 'success' of science was typically measured with quantitative indicators which fail to describe the quality of the research, the researchers' experience in their social contexts or their contribution to society (Schouten, 2019).

STRENGTHENING LOCAL RESEARCH CAPACITY

Capacity development and research in the water sector are essential for developing countries to meet their national sustainable development agendas as well as their international obligations such as the United Nations Sustainable Development Goals (SDGs). However, developing countries remain technically and financially under-resourced and are short on the evidential data and insight required to make informed decisions (McKenzie *et al.*, 2015; Almeshqab & Ustun, 2019). This exposes an implementation gap between the aspirations of the development goals and what is technically and financially feasible. International development organisations have often stepped in to fill gaps in research by importing expertise without considering enhancing the local capacity base in technical, institutional and financial sense on how to conduct, analyse and use the information collected; this base comprises the local researchers and the authorities and decision makers they are interacting with. Expertise, funding and ideas often flow from high-income countries which have less knowledge of the local, true needs (Beran *et al.*, 2017). In addition, the new expertise often stays within the international development organisation or its consultants/experts, having limited transfer to local researchers. Critically, the research for the evidence-based decision-making and the capacity development of local researchers are often treated as two independent tasks. The collection and analysis of evidence are conducted by the international experts, separate from any capacity development of local authorities and researchers (Wehn *et al.*, 2015). We propose that this need not be the case as alternative, better-integrated models exist in other disciplines, such as the health sector, which can be modified for the water sector. Research capacity building (RCB) is widely acknowledged in the health sector and has assisted in the career development of medical staff for more than 15 years (Lansang

& Dennis, 2004; Bates *et al.*, 2006). Although similar efforts have taken place in the water sector, they may be fewer or less generic and are less well documented. The RCB approach defines capacity development as a continuous process of empowering individuals, institutions and nations to (1) define and prioritise the problems they are facing in a systematic way; (2) scientifically evaluate appropriate solutions; and (3) apply and disseminate the generated knowledge (Lansang & Dennis, 2004).

This paper reviews a pilot on Capacity Building through Research (CBtR) on water. The pilot provides a platform to enhance scientific analytical skills, by facilitating leadership of local experts in the definition and implementation of projects, and including cross-sectoral collaboration where appropriate. This approach relies on a heightened level of critical thinking whereby local researchers are tasked with defining the objective and research questions, collecting and analysing the data, and supplying the resulting information to decision-makers. It adopts the concept that evidence provided by local experts who are, however, guided by international experts – who can also be perceived as impartial and knowledgeable – will benefit from better buy-in by local institutions and policymakers, which should result in solutions that are more sustainable and implementable than those supplied or imposed mostly by outsiders (Lansang & Dennis, 2004; Beran *et al.*, 2017).

Literature suggests that sustainable local knowledge systems are needed to support sustainable development. Hands-on learning through ongoing research programmes has been shown to be an effective learning mechanism which complements existing academic education (Lansang & Dennis, 2004). A CBtR approach enhances application skills, improving not only the ability to design and conduct research, but also to evaluate the results and use the results effectively (Bates *et al.*, 2006). Experiences from the health sector caution against 'bolting on' capacity development to existing research projects (Bates *et al.*, 2006), indicating that conscious design, planning and integration of the learning experience from the outset are important. Furthermore, Lansang & Dennis (2004) argue that key performance indicators should be agreed upon that go beyond superficial quantitative statistics to assess the research environment itself. This is especially relevant in water management, where capacity development and collaboration across sectors must be incorporated in programmes if they wish to contribute to sustainable solutions. Water management problems are multi-faceted and inter-disciplinary and implementation needs to be sensitive to this (Sheil *et al.*, 2016). For example, de Silva *et al.* (2017) report from the water (fisheries and agriculture) sector where local stakeholders were engaged across scales and sectors in order to define problems, research objectives and management options in line with wider local sustainable development goals; this very process was found to build consensus and capacity while reducing conflict. White *et al.* (2019) found that bringing together researchers, water managers and policy makers, amongst others, assists in the formation of research projects that are more appropriate, accepted and transformative in local contexts.

REVIEW OF CBTR PILOT PROJECTS

Building on work done in the health sector through RCB, the CBtR approach was piloted in several Dutch- and Australian-funded programmes and smaller individual projects, five of which are reviewed here, all located in Myanmar and taking place in 2015–2020. In these initiatives, the focus was on local practitioners and academics, and on building their technical skills through project-based research and training, in which international experts had only an assisting role. To assess the CBtR approach, the research question was formulated as to whether it enables local researchers to collect primary data for use in decision-making and increases their capacity to conduct their research and communicate results more effectively.

Study area

The pilot projects were located in Myanmar, a developing country in South East Asia. Myanmar's vast water resources are estimated to constitute up to 12% of the whole of Asia's freshwater resources and 16% of that of the ASEAN nations (Myint *et al.*, 2020). They comprise numerous large lakes, eight main river basins, and the Ayeyarwady mega-delta (Anthony *et al.*, 2019). Water availability is critical to the 54 million inhabitants as 70% of the population is rural, engaged in agriculture and reliant on natural resources (FAO, 2019). Inland fisheries are economically, nutritionally and socially significant (Duncan *et al.*, 2021). However, increasing pressure from irrigation and hydropower (Emmerton *et al.*, 2015) supporting Myanmar's growing population will make effective water management an urgent and essential challenge in meeting sustainable development goal objectives and improving livelihoods.

Project evaluation framework

The projects were analysed and assessed by adopting criteria from the recently issued Strategy Evaluation Protocol (SEP, 2020) which reflect the importance of institutional collaboration for institutional strengthening and development. The SEP acknowledges that until recently success criteria for scientific achievement were predominantly quantitative. However, these do not necessarily represent the main functions of science in society, which have historically been defined as the pursuit of knowledge with the highest standards of objectivity and the elimination of error through transparent, open and reproducible practices. Worldwide, organisations supporting research are adjusting their evaluation criteria by incorporating SEP to account for the capability of science for facilitating societal breakthroughs (Schouten, 2019). Notably, in the Dutch academic system, research quality and relevance to society are the two qualitative domains that are taken into consideration for assessing research. In the CBtR approach, this second qualitative domain is particularly pertinent for the evaluation of scientific achievement, and it was therefore selected as a key criterion in this study. For this domain, the SEP specifies the following sub-questions: (1) how the research product (such as guidelines, software and websites) is communicated and perceived by target societal groups or the general public (popular science); (2) how the research products are used by society, including the demonstrable uses of scientific results by social institutions, companies and governments, practitioners, teachers and media users, among others; and (3) how society recognises researchers, which includes the attribution of awards or means of support, important collaborative grants, public prizes and demonstrations of praise, and secondary appointments in civil-society organisations or policy-driven institutions. In this study, the CBtR pilot projects in Myanmar were analysed and evaluated for their contribution to evidence-based decision-making as well as for their inherent quality, based on the three SEP sub-questions: communication, uptake of research products and societal recognition of the research.

Pilot projects

In each of the five pilot projects described below, international experts from research institutes and/or small or medium enterprises would work with and mentor a local group (e.g. from government, NGO or an academic department) and/or an individual (e.g. researcher), and work together over a defined period on projects or academic curricula to define a series of research questions, develop research skills, conduct the research, analyse and share results with decision-makers, and incorporate the lessons into their professions or academic curricula.

(1) Water Land and Ecosystems (WLE) project (2015–2017) funded by CGIAR. The WLE project aimed to contribute to the sustainable and integrated development of Myanmar's water resources by enhancing capacities for inclusive governance and management of water and ecosystem services. A professional training and mentoring programme for a cross-sectoral group of mid-level water professionals was developed with three objectives: to create a common understanding of water, land and ecosystems in the Ayeyarwady river basin; to enhance personal research skills and to focus on active learning and coached project work. Between 2015 and 2017, a total of 37 Myanmar mid-level professionals participated in the respective programmes. Training was embedded in a policy engagement track focusing on the uptake of research and generating

understanding of the science–policy interface in the water governance and management arena. The policy engagement track further ensured a reflection on policy relevance during project implementation. In addition, in a separate track, institutional capacity was developed in key Myanmar organisations (universities but also the local river basin network) on water management and governance through discussions on IWRM certification and institutionalisation of the training. Special attention was given to communication of the research results, through pitch training, debating sessions and participation of fellows in national and international conferences. The WLE programme comprised a project development phase in which participants were tasked with developing a theory of change for a given project in their area of interest. The aim was to not only learn in class, but also to apply these concepts and methods on the ground in their current organisation. The objectives of the project phase were to better appreciate the need for the integration of concepts and challenges; to reflect on the project based on what was learned in class; and to make recommendations for their own organisation as well as for other organisations and the National Water Resources Committee. The project phase followed an eight-step approach which detailed the process from problem description to envisioned change in a number of clear monitorable steps. Discussion sessions and mentoring were carried out regarding the Myanmar National Water Policy as a framework and tool to be used in policy, planning and project development. Together, this would strongly encourage participants to connect societal and technical issues and identify pathways for change, but also build their capacity to articulate project proposals and plans and shift from 'safe' technical reports towards fundable and actionable plans for donors and institutions. An important component was the association with the three universities with a water-related curriculum, i.e. Yangon Technological University (YTU), Mandalay Technological University (MTU) and Myanmar Maritime University (MMU).

(2) NICHE Myanmar 250 project on Capacity Building for IWRM at two key universities, funded by the Dutch Organisation for Internationalisation of Education (NUFFIC). This project targeted the universities exclusively, however, the collaboration by WLE with this project further strengthened its relationship with the above universities. Apart from supporting universities in developing new curricula on IWRM, NICHE 250 had a dedicated component for staff development. This provided the space to develop a training-of-trainers programme for university staff participating in WLE with some additional coaching.

(3) S-MultiStor project (sustainable hydropower and multipurpose storage to meet water, food and energy development goals: a program for collaborative research and innovation), funded by the Netherlands Ministry of Foreign Affairs. This project aimed to strengthen local technical and scientific communities by jointly investigating and demonstrating approaches to render water storage more sustainable. Myanmar, specifically the Ayeyarwady Basin, was one of the three geographic focuses of this global project which also includes research in the Magdalena and Zambezi basins in Colombia and southern Africa, respectively. S-MultiStor was designed to support the engagement of partner universities in the global South in ongoing international initiatives related to sustainable hydropower and multi-purpose water storage, enriching those initiatives through research conducted by staff and students with support from international experts.

For Myanmar, two thematic areas included research development: to guide the implementation of the national Directive on environmental flows as affected by both hydropower and multi-purpose irrigation dams; and to improve fish migration through better management of multi-purpose dams and across river and wetland infrastructure. The project involved the collaboration of numerous partners including local scientific and technical partners and policy-makers (YTU, MMU and the Ministry of Agriculture, Livestock and Irrigation [MOALI], NGOs with local implementation experiences such as World Wildlife Fund (WWF) and Fauna and Flora International (FFI), international research institutes (IHE Delft, The Netherlands, and Charles Sturt University, Australia) and the International Finance Corporation (IFC, World Bank Group). Tangible outcomes of the project included the introduction of a new paradigm of system-scale planning tools and of basin-scale approaches in the local practices and in the research and education programmes of the academic institutions. This necessitated cross-sectoral involvement in flow release scheduling from dams, irrespective of dam ownership.

(4) Delta Flows project (on the role that deltas play in sustaining basin-scale fisheries in the Mekong and Ayeyarwady rivers), funded by the Australian Government and the Australian Water Partnership. This project sought to investigate the river infrastructure's impact on fish migration. At present, very little is known about migratory fish in Myanmar and local capacity is lacking to conduct research able to inform infrastructure-related decision-making (Conallin *et al.*, 2019). The project aimed to generate key knowledge on migratory species; strengthen research capabilities and cross-sectoral collaboration; and facilitate the application of that knowledge in policy and management practice. Local fisheries staff at FFI Myanmar and the Myanmar Department of Fisheries (DOF) were taught how to remove the ear bones (otoliths) from different species of fish that were thought to be migratory. Otoliths from a specific economically important fish called Hilsa (*Tenualosa ilisha*) were sent for laboratory analysis to investigate their migration between the sea and freshwater as it was suspected that sluice gates in Myanmar's river deltas were impacting Hilsa populations. With this information, the DOF would be able to discuss the operation and refurbishment of existing sluice gates with the Irrigation Department which owns and constructs the sluices so that sluice infrastructure in the future might not impede Hilsa migration.

(5) Leapfrogging Delta Management project – Showcasing Smart Information Solutions in the Ayeyarwady Delta, funded under the Dutch Partners for Water programme. This project aimed at bypassing less efficient and more expensive technologies and moving directly to state-of-the-art applications in measurement and monitoring technology taking full advantage of ICT. Such innovation involves a long process and not seldom trial and error. It proved possible to integrate it with the other CBtR projects. This served several purposes, as it sparked scientific and technological curiosity and required critical thinking and deep understanding of the monitoring variables; it also imposed strategic thinking on data management as many innovative technologies make use of mobile phone and cloud data services. Lastly, bringing together innovation entrepreneurs, developers, university staff and students, and end-users would help the innovators to improve their products for the local market.

ANALYSIS AND EVALUATION OF CBTR PILOT PROJECTS

To assess the five pilot projects regarding their effectiveness in strengthening local capacity for the collection and analysis of basic data relevant for decision-making, they were analysed, and a brief survey was conducted, applying criteria from the SEP. The different approaches in the above pilot projects yielded complementary information that, taken together, generated insight into both the learning process of Myanmar participants and the functioning of CBtR.

The S-MultiStor project allowed the CBtR approach to be applied systematically, especially through the collaboration with the academic partners. The concepts of multi-objective planning and system-scale planning are novel to Myanmar, and several PhD and Master-level study projects were supported to help introduce these concepts; they also contributed to updating the post-graduate curricula. Updates concerned training on the development and application of tools for system-scale and multi-objective planning, environmental flows assessment and recommendations, and the importance of connectivity in river basins for sustainable development. MSc and PhD fellows experimented with devising environmental flow recommendations for various river systems downstream of large multi-purpose dams to address ecosystem degradation caused by dam construction. Hydrology-based environmental flow models were piloted and adjusted by the students to suit local conditions and data availability

(Hlaing *et al.*, 2020). In addition to these modelling exercises, several field trips were conducted by YTU and MMU to collect hydrological, water quality and social data from the studied systems, and for tool refinement. Besides the completion of several MSc theses, a PhD study was recently defended, supported by S-MultiStor, on the hydropower potential in the Myitnge River basin which introduced hydropower investment sustainability as an impact analysis criterion; such analysis had not yet been considered in similar studies in Myanmar. In the framework of this research, the student took an internship at IHE Delft, The Netherlands, for collaborative work with this and other Dutch research institutions. The topics of river basin connectivity and environmental flows are also now embedded in the local curricula and in student projects.

In parallel, an environmental flows group was established in civil society to investigate the release of environmental flows from both hydropower and multi-purpose irrigation dams. This involved a core group of university researchers (YTU, MTU and MMU), NGOs (WWF) and the IFC. This group focused on three points to strengthen institutionalisation and implementation of environmental flows: policy and legislation; data and research; and institutional roles and capacity. The workshop Exploring the Science and Policy for Developing a National Environmental Flows (eflows) Framework for Myanmar was held in Yangon in 2018, and the Symposium and Technical Workshop on Environmental Flows in Myanmar in 2019 was co-organised by IHE Delft, Charles Sturt University, YTU, the Australian Government, Badu, Riversfutures, WWF Myanmar and the Australian company Alluvium. Both initiatives were attended by over 100 experts from academic institutions, engineering companies, various stakeholders and government agencies. In 2018, a technical workshop with 60 local and international experts on environmental flows helped set a baseline for devising a national framework. A state-of-knowledge review on environmental flows within Myanmar was also published.

A second civil-society focus group discussed migratory fish and multi-purpose dams, and the impact of river and wetland infrastructure on fish migration. This group has collaborated with MOALI and DOF on the options for improving fish migration through infrastructure improvement. Since 2018, cross-institutional World Fish Migration Day events have been held in Myanmar and the Netherlands, which included outreach initiatives in both countries, a 'wicked debate' in the Netherlands and panel discussions in Myanmar (in Yangon, Lake Indawgyi and online). These events debated improving awareness of the impact of water infrastructure on fish migration and the resulting challenges for biodiversity, wetland ecosystems and food security.

A simplified CBtR approach was applied successfully in the Delta Flows project. Key fisheries and NGO staff were trained. They accurately and systematically collected and processed a sufficient number of otoliths for investigating migration. Results demonstrated migration patterns for the Hilsa between the sea and the rivers. The DOF and FFI shared this information within their own institutions and also in universities to improve knowledge of the critical importance of fish migration. DOF discussed with the Irrigation Department on how Hilsa and other fish species are being affected by sluices and other water infrastructures. The information on Hilsa migration routes is also being utilised by another (on-going) donor-funded project aimed at increasing fish passage and migration at key barriers.

The introduction of innovative water management measurement and monitoring technology within a CBtR framework was broadly perceived as successful from both an economic and a capacity development perspective. The universities have incorporated the innovations into the curricula and practical assignments. The novel monitoring and measurement technology is now being used routinely in MSc and PhD research. Staff was trained in the techniques to instruct local users who have now gained independence from external experts. As an example, water quality assessment, river bathymetry, wave monitoring and suspended sediment transport in rivers have become standard elements in YTU and MMU curricula. Finally, a local engineering company specialised in IWRM and water management has been established by trainees of the programme.

To gauge how WLE project participants had experienced the CBtR approach and to what extent they would apply the new knowledge in their future career, a brief survey was conducted among the participants featuring questions derived from the SEP evaluation criteria as outlined above. Table 1 provides a set of qualitative but representative responses by respondents; the limitations in participant sub-group numbers and the diversity in their background rendered a more rigorous statistical analysis less meaningful however.

The results show that, generally, participants were able to value and articulate the benefits of the projects' focus on CBtR from design to evaluation and communication. On the other hand, when asked to reflect on the impact of the training on their future professional career, few responses were volunteered. Similarly, it appeared that the most difficult question to respond to concerned the advice they would give to seniors in their organisation, and their policy recommendation for the river basin in general. This restraint was likely due to the hierarchical inter-personal relationships in Myanmar (which discourages juniors to directly advise seniors) as the projects involved primarily mid-level professionals. Nonetheless, some participants demonstrated knowledge on how they would apply the acquired capacity in their job, either by training other professionals and creating a multiplier effect (which was more obvious for the university participants as they develop curricula) or by furthering the prepared project proposals in their respective organisations.

DISCUSSION

The CBtR pilot projects demonstrate the importance of institutional capacity both to ensure proper implementation of the projects themselves and to lay the foundation for further replication of the capacity development. CBtR encouraged effective project definition and implementation, because its research expanded scarce information

Table 1 | Evaluation of the WLE project, typical representative responses.

Question posed	Response
What is the most significant knowledge you gained during the training program?	IWRM, governance, interconnection of problems, need to collaborate, proposal writing
What is the most useful skill you acquired during the training?	Debating, presentation skills (5-min pitch), listening, proposal writing
How has your network grown during the training and how have you used it in your research/proposal development and further work?	Network of friends from different disciplines, organisations, exchange of data to make proposal, also skills, international network and training opportunities (one fellow presented proposal to RD and wants to do a project)
What is the key message you had for your senior/colleagues after completing the training?	No response
How do you think positive change can be achieved in water management in Ayeyarwady? What would be your advice/ policy recommendation after completing your research/ proposal?	No response
How were you able to use in your job the knowledge and connections that you acquired in the training?	Staff who participated in training-of-trainers programme now use the skills and content in their teaching (multiplier effect), knowledge immediately translated in viable project (YCDC)

in a data-poor environment which in addition was designed so as to generate information which would be directly useful for decision-making on pending questions. In addition, attention to developing capacity in communication by skills training such as pitching, debating and presenting remains important. In one example, data collected as part of S-MultiStor and presented by local university researchers helped inform WWF to allocate funding for project continuation. Taking into account the current pressure in the country to build new hydropower facilities, the sharing of knowledge and collaboration are essential to strengthen local governance and assist agencies in making more sustainable choices for water management. Each of the cases demonstrates the pertinence of the second criterion of the SEP's qualitative evaluation criteria, namely, to what extent the research outputs support their *use* by society. The WLE targeted technical staff at Myanmar's water sector organisations, many of whom were educated at YTU, MTU or MMU, making collaboration with those universities a logical starting point for developing capacity. The more academically oriented S-MultiStor programme translated this criterion into renewal of academic curricula at the post-graduate level. This will in turn contribute to achieving the first SEP criterion which pertains to the way scientific information is received by or communicated to (local) society. The inclusion of the new concepts of multi-objective and system-scale planning in water management training confirms the value of this information to Myanmar's local context; it was incorporated in national course materials to become available for further development to address local and national questions and be useful to actors across academic, professional and governance spheres. Interestingly, the environmental flows focus group brings together members from academic, policy, financing and civil society groups, encouraging collaboration and enabling each of them to utilise the enhanced capacity for identifying and addressing water management challenges.

To enhance the effectiveness of CBtR and project-based learning, the presence of trainers on the ground and local fieldwork were found to be crucial. Participants responded best to coaching on topics that were of clear mutual interest. Admittedly, this format is resource-intensive, but the potential for remote coaching proved limited even when relationships had been already established. Locally active organisations such as NGOs can be crucial partners in facilitating a CBtR approach when international experts are temporarily absent, avoiding discontinuous project implementation (Brown, 1991). Such bridging organisations can be effective in linking international actors with local practitioners and communities (Sapsed *et al.*, 2007). They strengthen project legitimacy by building on their existing rapport with the community; working from existing social networks also tends to be less time-consuming and expensive than starting a top-down initiative from scratch. These factors help conduct relationship-based, long-term and outcome-focused capacity development, which generally requires a commitment beyond the usual government cycles (Prager, 2015).

Feeding back acquired new capacity into the participants' organisations posed specific challenges. This was to some extent related to an equivocal commitment from their superiors and organisations to put the new capacity into practice. Although this commitment was explicitly or implicitly requested at the start of the pilot projects, the communication and selection procedure of candidates did not sufficiently emphasise this to ensure that the organisations would be prepared to incorporate the acquired capacity. This reluctance is likely also a reflection of the academic and organisational cultures in the country with their strong hierarchical behaviour that tends to discourage independent contributions from younger or lower-level staff. Champions in senior staff and trust-building can be instrumental to make new ideas enter into existing workflows; typically, this is a process that exceeds the typical lifespan of a CBtR project. Finally, networks surrounding the CBtR help establish a community of users in the country. Here, this was achieved by involving representatives of universities and key water agencies such as the National Water Resources Committee in the training programmes and by helping participants develop their local and international networks through participation in water events and social networks.

To assure the continued success of CBtR requires a long-term commitment from both government and donor agencies to ensure that an effective knowledge base and adequate research capacity are built and that pathways are established for the application of the generated data and knowledge. Support programmes should be aware that building local capacity demands substantial investment and resources and that it should include an assessment of the potential and commitment of local partners to facilitate learning through project implementation. Indeed, the pilot projects suggest that the full benefits of CBtR can be captured only over a time span that is generally much longer than the programme's duration or its funding or political cycles; this requires broad-based and cross-sectoral support.

CONCLUSIONS

This paper reports the initial observations of five pilot projects in Myanmar applying an approach which combines data collection and research implementation, with the structural strengthening of institutional capacity to conduct research (CBtR). As the overview shows, a major weakness in current international development is that projects often are comparatively small, have a short duration and are poorly connected to similar local and international initiatives. While some projects do aim at developing research capacity, they tend to have modest ambition and short-time horizons; many other projects, however, tend to aim solely at quick, practical results in terms of data and information collection to address isolated solutions. Projects are often driven by the specific technical interests and budget situation of the funding agency rather than by broader systemic interest in institutional capacity development. The observations suggest that knowledge that is developed locally through project-oriented research in which local and international experts work, explore and learn together leads to better informed and more evidence-based decision-making. As a first benefit, participants from different levels of government demonstrated greater awareness and understanding of the issues they are facing and which need to be addressed in order to meet sustainability goals. In Myanmar, such action can be taken within the National Water Policy Framework which lends it a greater legitimacy. The proper design of projects allows to effectively combine research implementation and research capacity development, instead of approaching these as two separate goals along different avenues. Still, several challenges exist to the implementation of CBtR in developing countries, in part because project impacts are cumulative. To allow long-term change to take root, a critical combination of influencing factors and decisions is needed. This requires considerable and sustained investment in time and funding; however, such long-term commitments may be less compatible with the short-term nature of many international development approaches, and starting new initiatives with different objectives overlaying old ones can be confusing and counterproductive. Therefore, projects are preferably designed with clear short- and long-term goals. These will allow to keep generating deliverables and benefits that can persuade funding and oversight institutions, but at the same time develop the endogenous (research) and local institutional capacity which will mitigate the risk of discontinuation due to changes in institutional or other structural support. A key conclusion of the study is that CBtR does strengthen the role of local universities as knowledge institutes capable of initiating and conducting research more autonomously and producing evidence for informed decision-making by sector agencies and communities. In line with the SEP criteria, this is likely to lead to better sharing and use of scientific information which in turn will fundamentally contribute to the capability of developing countries to understand and manage their sustainable development challenges.

DATA AVAILABILITY STATEMENT

All relevant data are included in the paper or its Supplementary Information.

REFERENCES

Almeshqab, F. & Ustun, T. S. (2019). Lessons learned from rural electrification initiatives in developing countries: insights for technical, social, financial and public policy aspects. *Renewable and Sustainable Energy Reviews 102*, 35–53. https://doi.org/10.1016/j.rser.2018.11.035.

Anthony, E. J., Besset, M., Dussouillez, P., Goichot, M. & Loisel, H. (2019). Overview of the monsoon-influenced Ayeyarwady River delta, and delta shoreline mobility in response to changing fluvial sediment supply. *Marine Geology 417*, 106038.

Bates, I., Akoto, A. Y. O., Ansong, D., Karikari, P., Bedu-Addo, G., Critchley, J., Agbenyega, T. & Nsiah-Asare, A. (2006). Evaluating health research capacity building: an evidence-based tool. *PLoS Medicine 3*(8), e299. https://doi.org/10.1371/journal.pmed.0030299.

Benson, D., Gain, A. K. & Rouillard, J. J. (2015). Water governance in a comparative perspective: from IWRM to a 'nexus' approach? *Water Alternatives 8*(1), 756–773.

Beran, D., Byass, P., Gbakima, A., Kahn, K., Sankoh, O., Tollman, S., Witham, M. & Davies, J. (2017). Research capacity building – obligations for global health partners. *The Lancet Global Health 5*(6), e567–e568. https://doi.org/10.1016/S2214-109X(17)30180-8.

Biswas, A. K. (2004). Integrated water resources management: a reassessment. *Water International 29*, 248–256.

Brown, D. (1991). Bridging organizations and sustainable development. *Human Relations 44*(8), 807–831. https://doi.org/10.1177/001872679104400804.

Conallin, J. C., Baumgartner, L. J., Lunn, Z., Akester, M., Win, N., Tun, N. N., Nyunt, M. M. M., Swe, A. M., Chan, N. & Cowx, I. G. (2019). Migratory fishes in Myanmar rivers and wetlands: challenges for sustainable development between irrigation water control infrastructure and sustainable inland capture fisheries. *Marine and Freshwater Research 70*, 1241–1253.

de Silva, S., Miratori, K., Bastakoti, R. C. & Ratner, B. D. (2017). Collective action and governance challenges in Tonle Sap Lake, Cambodia. In *Water Governance and Collective Action: Multi-Scale Challenges*. Suhardiman, D., Nicol, A. & Mapedza, E. (eds). Routledge, London, pp. 108–119.

Duncan, N., de Silva, S., Conallin, J., Freed, S., Akester, M., Baumgartner, L., McCartney, M., Dubois, M. & Senaratna-Sellamuttu, S. (2021). Fish for whom?: Integrating the management of social complexities into technical investments for inclusive, multi-functional irrigation. *World Development Perspectives 22*, 100318. https://doi.org/10.1016/j.wdp.2021.100318.

Emmerton, M., Thorncraft, S., Oksanen, S., Soe, M., Hlaing, K. K., Thein, Y. Y. & Khin, M. (2015). *Myanmar Energy Master Plan*. National Energy Management Committee, Myanmar. Available at: https://www.burmalibrary.org/docs22/2015-12-Myanmar_Energy_Master_Plan.pdf.

FAO (2019). *Country Profiles: The Republic of the Union of Myanmar*. United Nations Food and Agriculture Organisation, Rome. Available at: http://www.fao.org/fishery/facp/MMR/en.

Grigg, N. S. (2016). *Integrated Water Resource Management: An Interdisciplinary Approach*. Palgrave MacMillan, London.

Hlaing, S. S., Zin, W. W., Thin Kyi, C. C. & San, Z. M. L. T. (2020). Hydropower generation considering the environmental flow in Myitnge river basin, Myanmar. In *River Flow 2020* (W. J. Uijttewaal, M. Franca, D. Valero, V. Chavarrias, C. Ylla Arbós, R. Schielen, R. & A. Crosato, eds.) CRC Press, London, pp. 2381–2387.

Lansang, M. A. & Dennis, R. (2004). Building capacity in health research in the developing world. *Bulletin of the World Health Organization 82*, 764–770.

McKenzie, F. C., Williams, J., Daugbjerg, C., Grafton, Q. & Qureshi, E. (2015). Sustainable food production: constraints, challenges and choices by 2050. *Food Security 7*, 221–233. https://doi.org/10.1007/s12571-015-0441-1.

Myint, K., Wint, T., Naing, K. M., Win, S. S., Soe, M. M., Oo, T., Maw, Y. Y., Swe, E. E., Ni, D. N., Lwin, T., Shwe, W. W., Myo, W. P. P. & Aung, K. W. (2020). *Water Resource Management in Myanmar [Course Notes]*. Open University. Available at: https://www.open.edu/openlearncreate/mod/oucontent/view.php?id=156220.

Prager, K. (2015). Agri-environmental collaboratives as bridging organisations in landscape management. *Journal of Environmental Management 161*, 375–384. https://doi.org/10.1016/j.jenvman.2015.07.027.

Sapsed, J., Grantham, A. & DeFillippi, R. (2007). A bridge over troubled waters: bridging organisations and entrepreneurial opportunities in emerging sectors. *Research Policy 36*(9), 1314–1334. https://doi.org/10.1016/j.respol.2007.05.003.

Schouten, J. (2019). Connecting science and society. *Impact* (2), 44–45. https://doi.org/10.21820/23987073.2019.6.58.

SEP (2020). *The Strategy Evaluation Protocol 2021-2027*. VSNU, KNAW and NOW, The Hague, p. 48.

Sheil, C., Leal Filho, W., do Paço, A. & Brandli, L. (2016). Evaluating the engagement of universities in capacity building for sustainable development in local communities. *Evaluation and Program Planning 54*, 123–134. https://doi.org/10.1016/j.evalprogplan.2015.07.006.

Wehn, U., Irvine, K., Jaspers, F., Douven, W., Pathirana, A. & de Ruyter, E. (2015). Strengthening water governance in the global south: role and international experiences of UNESCO-IHE in capacity development. *Water Governance 05*, 26–34.

White, D. D., Lawless, K. L., Vivoni, E. R., Mascaro, G., Pahle, R., Kumar, I., Coli, P., Castillo, R. M., Moreda, F. & Asfora, M. (2019). Co-Producing interdisciplinary knowledge and action for sustainable water governance: lessons from the development of a water resources decision support system in Pernambuco, Brazil. *Global Challenges* 3(4). https://doi.org/10.1002/gch2.201800012.

First received 23 December 2020; accepted in revised form 3 February 2022. Available online 3 March 2022

© 2022 The Authors From Knowledge and Capacity Development to an Implementation Science: doi: 10.2166/wp.2021.094

Contextual knowledge co-production and capacity building for sanitation planning: experience from Kerala, India

Sruthi Pillai* and N. C. Narayanan
Centre for Technology Alternatives for Rural Areas, Indian Institute of Technology Bombay, Powai, Mumbai 400076, India
*Corresponding author. E-mail: sruthipillai1234@gmail.com

SP, 0000-0002-0561-8198

ABSTRACT

Small towns in the global South face huge infrastructure deficits in sanitation due to the lack of technical, institutional and financial capacity to establish treatment systems. This knowledge gap and capacity deficit are a socio-ecological challenge that requires knowledge co-production by both science and society. This work focuses on a transdisciplinary initiative on capacity building and knowledge co-production called CANALPY in Alleppey, a town located in the southern state of Kerala, India. The project recognises the need for data and knowledge for developing a fine-grained understanding of the sanitation situation in the town and also raises awareness about canal pollution to arrive at participatory solutions. This study analyses the drivers of co-production of knowledge with its enabling conditions as well as the challenges associated with it. The initiative has been instrumental in creating a space and platform for collaborations, dialogues and discussions on problems of sanitation, water quality and pollution. CANALPY has ensured the dissemination of findings among a larger audience and has inspired new thinking and action. It is also important to take into critical consideration the different practical, socio-political, institutional and competence-related challenges that impede the project's goal of knowledge sharing and capacity building.

Key words: Capacity building, Knowledge co-production, Sanitation, Transdisciplinary research

HIGHLIGHTS

- Capacity deficit in urban local bodies impedes sanitation service delivery, exacerbating environmental challenges.
- The study explores CANALPY, a transdisciplinary initiative to address capacity and knowledge deficit.
- The project led to the creation of a knowledge repository on the town and mechanisms for transferring them to decision makers.
- Understanding enablers and challenges of the CANALPY initiative is critical to adapting the experience to other small towns.

INTRODUCTION

Urban areas all over the globe are facing social and environmental challenges related to climate change, resource scarcity and inequality. Most of the tier two and three cities are already struggling to deliver infrastructure and services due to limited institutional capacities and finances (Brown *et al.*, 2012). Small towns are plagued by substandard sanitation infrastructure and services due to governance constraints, paucity of resources and capacity deficit of local governments (Chong *et al.*, 2016) leading to contamination of open wells and consequently shallow aquifers (Bhosale & Kumar, 2002; Quamar *et al.*, 2018) affecting water quality and public health. Though large cities can afford conventional solutions like Sewage Treatment Plants, small towns are faced with huge infrastructure deficits due to lack of technical, institutional and financial capacity (Tutusasus & Schwartz, 2018). The limited investments in water and sanitation infrastructure in developing countries including India are ineffective due to the lack of operation and maintenance capacities (Subramanyam, 2020).

Addressing these concerns will require new skills, knowledge and competencies at the levels of household, community and local governments (Sharma *et al.*, 2014). Capacity deficits at the level of local governments have been identified with respect to climate change adaptation in the Philippines (Porio, 2011) and policy formulation and implementation in Delhi (Hughes, 2013). Urban local bodies (ULBs) in India do not have the technical, managerial and financial capacity to undertake water and sanitation-related projects (Rosenqvist *et al.*, 2016; Narayanan *et al.*, 2018b). This crisis of knowledge gap and capacity deficit is a socio-ecological challenge which requires knowledge co-creation and co-production by both science and society (Jagannathan *et al.*, 2020).

Researchers from multiple disciplines need to collaborate with each other as well as stakeholders from the state and civil society to address problems which emerge in real-world scenarios (Rozance *et al.*, 2020; Croese, 2021). Such an approach of integration of knowledge and methods to address challenges at the interface of scientific questions and societal problems is known as trans-disciplinary research (Jahn *et al.*, 2012). This approach is also credited with increasing citizen empowerment and ownership, a lack of which is indicative of a decrease in public trust in new technologies (Fritz & Binder, 2018). However, transdisciplinary research has mainly emerged in the global north where a legitimate and formalised structure is in place so that researchers from universities can engage with the institutionalised stakeholders (van Breda & Swilling, 2019). This is in stark contrast to the situation in the global south where complex, heterogeneous and socially stratified urban systems result in a lack of institutional spaces where formal stakeholders can engage with scientific expertise or state representatives (van Breda & Swilling, 2019). This study focuses on one such transdisciplinary initiative of capacity building and knowledge co-production efforts underway in the town of Alleppey located in the southern state of Kerala in India.

CANALPY[1] deals with knowledge production and transfer between the academic institutions (Indian Institute of Technology (IIT) Bombay), government agency (Kerala Institute of Local Administration, KILA), the local self-government of Alleppey Municipality and the community. The aim of this study is to critically analyse CANALPY to gain a deeper understanding of the enabling conditions or drivers of co-production of knowledge, the challenges and the outcomes. This study is also an attempt at mid-course reflective evaluation to improve the understanding of the team members about the status of the initiative and is expected to carve a roadmap to understand local civic governance challenges related to sanitation faced by small towns.

The next section describes the problems of sanitation and the need for interventions in small towns, followed by the case study site and a description of the activities of CANALPY. After detailing different activities of the initiative, the research methods employed are elaborated. This is followed by the findings of the empirical work detailing various factors which served as enablers and barriers for the initiative. Finally, we discuss the limitations of the work and the conclusion.

Problem of sanitation and the ongoing experiment of the CANALPY initiative

Alleppey, a heritage town in the southern state of Kerala, India (Figures 1 and 2), is renowned for its canals and backwaters. The canals have become progressively polluted over the last few decades due to indiscriminate dumping of solid and liquid wastes (Alappuzha Municipality & Government of Kerala, 2014; Karlaganis & Narayanan, 2014). This is linked to the crisis of sanitation which many of the towns in the global south currently face. In the

[1] Canalpy was initially conceived as a social media campaign; however, over the course of time, the name became synonymous with the whole initiative (https://canalpy.com/).

Fig. 1. | (a) Map of Alappuzha District (*Source*: GoK, 2014). (b) Location of Alappuzha, Kerala (*Source*: Authors).

face of challenges like the absence of quality data, low budgetary allocations and limited technical capacity, the design of any effective intervention will need a situation analysis of the state of sanitation to explore possible contextual solutions for the town. This was the motivation for the project between the Indian Institute of Technology, Bombay, Kerala Institute of Local Administration (KILA) and Alleppey Municipality.

The aim of the CANALPY initiative was to rejuvenate the heavily polluted canals, and the IIT Bombay team proposed a systematic situational analysis for understanding the baseline condition of the liquid waste management system at the town level. Earlier studies (Narayanan *et al.*, 2018b) proposed a contextual, participatory bottom-up sanitary mapping method which captures the social and spatial arrangements of small towns. The findings of the study demonstrated the need for reliable data for assessing the water and sanitation situation in small towns. Furthermore, the study indicates that government departments lack the capacity and resources to generate fine-grained data (Narayanan *et al.*, 2018b), and it proposed having academic institutions as knowledge producers with students doing data collection and analysis (Narayanan *et al.*, 2018b). The CANALPY initiative sought to replicate the same strategy of engaging the community and academic institutions to develop local capacity.

Researchers from IIT Bombay formed the core research team with the support of the KILA. The initial batch of local volunteers were trained by the team from IIT Bombay and together formed the CANALPY team. The first activity of the CANALPY initiative was a winter school which aimed at conducting a preliminary situational understanding of the sanitation context of the town. After consultations with local youth, civil society organisations like Kerala Shastra Sahitya Parishad (KSSP) and local engineering colleges, a protocol using ODK Collect and OSM Tracker to collect data and track GIS features was developed. Through a set of exercises

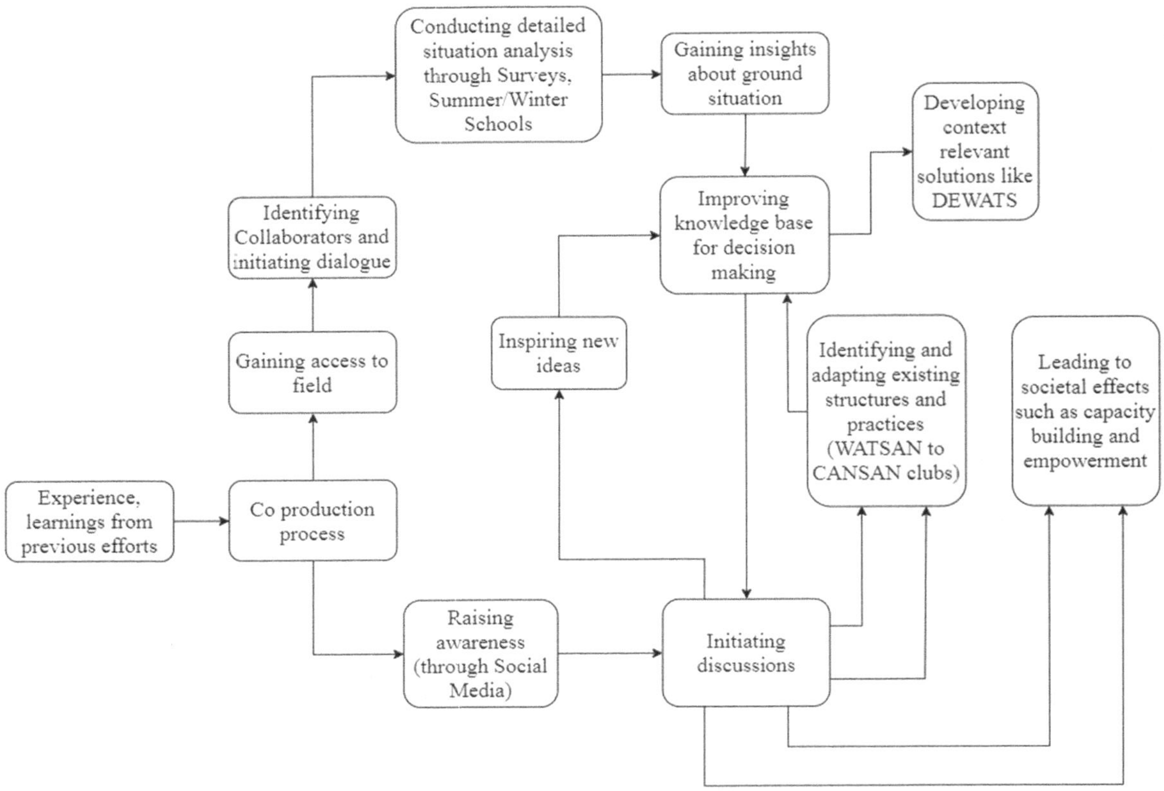

Fig. 2. | Illustration of the knowledge co-production process.

involving participatory drain mapping, water quality assessment and household surveys, the winter school findings conclusively proved the presence of sewage in the canals and prompted the need for a detailed study to understand the sanitation practices of the town (Narayanan *et al.*, 2017a, 2017b).

This was followed by Summer School 2018 which was a scaled-up effort supported by a wide range of participants from both technical and social studies backgrounds who conducted exercises like socio-economic and sanitation practices surveys, willingness to pay surveys, civil surveys of the canals and sub-canals, and studies of bacteriological contamination of water samples. The participants from around the country collected data from 1,500 households and 221 commercial/industrial establishments. The analysis and visualisation of the collected data helped in identifying the major pollution hotspots, polluters and highlighted the disparity in access to services for different socio-economic categories at the ward level.

After analysis of the collected data, canal sheds based on the concept of wastewater sheds (Narayanan *et al.*, 2018b) were made the basic unit of planning and intervention. The data collected during the surveys reflect the social and spatial aspects of pollution and the sanitary practices in the town leading to the demarcation of 'sanitation zones'. Sanitation zones can aid planners in identifying different typologies of sanitation practices and in developing context-specific interventions for each of the localities. The third activity was a winter school in December 2018 which aimed at creating canal shed-specific sanitation plans. The results and insights drawn from each of the studies were shared with the local government, Departments of Irrigation and the Pollution

Control Board along with recommendations for arriving at context-specific interventions for further action. One such context-specific intervention by CANALPY was the decentralised wastewater treatment system (DEWATS) for the most marginalised colony (Municipal Colony) of residents who belonged to the erstwhile manual scavenging occupation in 2019. The operation and maintenance of the DEWATS system was handed over to the members of the Municipal Colony in 2020. Similarly, on the basis of collected data, the Municipality and Kerala State Pollution Control Board conducted meetings with traders and citizen groups like Resident's Associations to encourage them to comply with environmental regulations followed by issuance of notices in case of violations. Timeline, as well as other activities of the initiative, is detailed in Table 1.

Figure 2 explains the nonlinear process flow which is the characteristic of a transdisciplinary action research project. Knowledge production activities are followed by knowledge exchange, like CANSAN, street plays, workshops and Massive Online Open Courses (MOOCs), which leads to the generation of new ideas that then contribute to sharpening original inferences. The ideation process is thus not limited to the researchers at the initial stages but is an iterative process with contributions from many of the collaborators and stakeholders involved in the process.

RESEARCH METHODS

The objective of the CANALPY initiative was to rejuvenate the canals through a participatory process of co-producing knowledge and design action through various activities. This explorative qualitative study deliberates on the drivers and barriers to the knowledge co-production process and attempts to evaluate the impact of the initiative and its allied activities until 2020. This study undertakes a case study approach which entails intensive and detailed analysis of a single case (Bryman, 2012). The authors were team members involved in the initiative, and the data collection involved multiple processes like key stakeholder interviews, focus group discussions (FGDs) and participant observations.

Table 1 | Timeline of CANALPY activities.

2015–2016	Previous research by team on wastewater sheds and sanitation zones. The team developed an approach to sanitation planning that leverages citizen engagement through participation and local capacities through collaboration with local governments (Narayanan *et al.*, 2018b).
2017	Team approached by Alappuzha Member of Legislative Assembly (MLA) Assembly to clean and rejuvenate the canals. Preliminary visit to town revealed solid and liquid waste pollution due to improper sanitation practices followed by discussions with ULB, MLA First winter school held in December.
2018	Series of surveys and studies conducted on waste and sanitation Summer school was conducted in May. Engagement with community initiated through institution building activities and workshops for local college students. Winter School 2018 for town-wide design of sanitation solutions.
2019	Clean-up organised by youth and community members of the town in one of the canals. CANSAN Campaign – aimed at sensitising school students initiated with the collaboration of education department and local municipality. Awareness building through a series of workshops and classroom sessions. Trained students further tried to build awareness in their communities through street plays.
2020	MOOC on 'Sustainable and Participatory Sanitation Planning' – course to replicate the Alleppey experience to other ULBs in Kerala, designed and delivered for elected members and officials of ULBs.

The research process started with key informant interviews who were either in positions of responsibility like elected representatives, officials and particularly those who were actively engaged in the activities of the CANALPY, and this led to an initial list of key informants. During an interview with key informants, they were asked to identify stakeholders not identified in the initial list (snowball sampling). In all, 24 stakeholders were identified and interviewed, and details are synthesised in Table 2. Stakeholders for the FGDs were also identified through purposive and snowball sampling. Two FGDs were held with the women from Municipal Colony.

The interviews followed a semi-structured approach. The questions were on the motivation, challenges encountered and the perceived outcomes of the initiative. FGDs aided in understanding the social dynamics of the community members, specifically the caste, class and gender relations. The interactions mapped the perception of the community members towards CANALPY including their concerns and conflicts related to the activities of the initiative.

Additional data were collected in the form of government reports, documents, publicly available information from the internet and field notes from participant observations in meetings and events. These observations aided in understanding the social relations, information flow and the decision-making process. The collection of data from multiple sources allowed corroboration of findings and identifying similarities and differences between the perceptions of different stakeholders. Data from all these sources were also triangulated to improve the reliability and external validity of the study.

The interview data and FGDs were translated from Malayalam to English and transcribed. While going through the various types of data collected, a constant comparative analysis was done to identify common and divergent themes. The text was then coded in Microsoft Excel according to the identified themes which allowed for categorising and analysing data with respect to the analytical framework identified in the following section.

Analytical framework

The findings from the initial set of activities of the initiative indicate that the problem of polluted canals of Alleppey is complicated by multiple pollution sources, multiple actors, capacity and resource deficit, ineffective monitoring mechanisms, and lack of public demand with the absence of accountability (Narayanan *et al.*, 2017a, 2017b, 2018a; Chhajed-Picha & Narayanan, 2021). This makes it a multi-actor, multi-scalar, dynamic, uncertain and unclear problem (Smith & Porter, 2010). Such a multi-causal socially complex problem involving multiple stakeholders is referred to as wicked problems (Peters, 2015). Local action through the activities of CANALPY has led to a certain amount of knowledge production, transfer and limited capacity building. Though the initiative is local, it is embedded in the broader institutional and societal context. To understand it, an analytical framework adapted from local management action by Patterson *et al.* (2013), as well as the work on collaborative decisions for environmental management by Margerum (2008), is proposed here (Figure 3).

FINDINGS

The findings section has been organised into two sections, the first examining the drivers for knowledge co-production and the second detailing the barriers and challenges to the co-production process.

Table 2 | Details of interviewees and stakeholder groups.

Stakeholder group	Number of interviewees
Government agencies at the state level	9
Community stakeholders	9
CANALPY team members	6

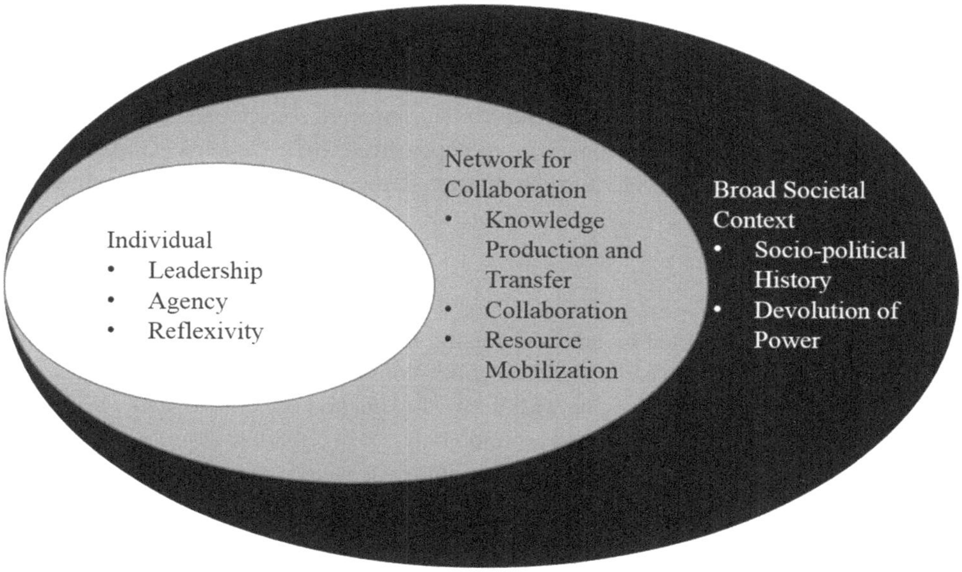

Fig. 3. | Analytical framework for enabling factors which served as drivers for the knowledge co-production process.

Drivers for co-production

Broad societal context

The broad societal context in the framework includes the following aspects: the socio-political history at the sub-national level and the devolution of power by the state government to the local self-government.

Kerala is widely known for its peculiar pattern of development, characterised by high levels of social development. The 'Kerala Model' as characterised by sympathisers includes a set of high material quality-of-life indicators coinciding with low per capita incomes, which was brought about by a set of wealth and resource redistribution programmes, investments in human resource development and progressive political participation (Dreze & Sen, 1997). The redistribution of assets by land reforms and the various pro-poor welfare policies were also the result of such politicisation, which led to the development of a welfare state, bringing better living and working conditions to large sections of people (Franke, 2019). The major criticism against the Kerala development experience is the slower rate of growth compared to even the Indian national average (George, 1998), the main reason being the lack of efficiency of the productive sectors of the economy. In addition to these, the environmental damage now faced by the state raises questions on the sustainability of even the limited economic growth achieved so far (Narayanan, 2003).

The 73rd constitutional amendment passed in 1993 in India granted new powers to the local government making them more accountable to the citizens. This was best operationalised in Kerala through the 'People's Campaign for Ninth Plan' from 1996 wherein local governments were given new functions and discretionary budgeting authority over 35–40% of the state's developmental expenditure (Heller *et al.*, 2007). Examining the socio-political history of the state reveals that democratic decentralisation ably assisted by grassroots democratic organisations has led to the strengthening of local institutions and planning capacity (Heller, 2001). Civil society organisations like the KSSP help mobilise people and their past experience of executing local-level initiatives and planning experiments including environmental accounting initiatives like the resource mapping was adopted by the Kerala State Planning Board (Heller *et al.*, 2007). The historical and political characteristics of Kerala play a central role in enabling knowledge co-production – the preoccupation with knowledge, education and literacy; broad-based democratisation and social change and concern with sustainability (Shrum & Ramanathaiyer, 2000). It is also important to understand the political and social context of Alleppey with high levels of literacy[2] and political engagement (Isaac, 1985). There is also a distinct environmental consciousness among the residents, evidenced by the decentralised sanitation campaign in Alleppey from 2012 (Isaac & Gopakumar, 2016). This environmental consciousness was also evident in the interest shown by community members who participated in the activities of CANALPY.

Network for collaboration

At this level, factors that were identified as enablers for the initiative were the process of knowledge co-production and transfer, collaboration between different organisations and resource mobilisation.

Knowledge co-production focuses on engaging with both scientific and societal actors by bridging disciplinary boundaries and solving real-life problems on the ground (Polk, 2014). Similar approaches are given different names such as interactive social research (Talwar *et al.*, 2011), transformative or participatory sustainability science (Lang *et al.*, 2012) and action research or participatory research (Wiek *et al.*, 2014). Some of the features of this type of approach are that (1) they facilitate collaboration among scientists and non-scientific actors, (2) decision makers and the communities strive to generate relevant and context-appropriate knowledge, (3) encourage participants to not only divulge information but engage in deliberation and reflection, (4) keep normative aspects like values central in the research process and finally (5) encourage mutual accountability, ownership and leadership among the participants (Talwar *et al.*, 2011; Lang *et al.*, 2012; Wiek *et al.*, 2014). The aim of this approach is to capture the complexity of societal problems by using knowledge from multiple disciplines and employ a context-specific, collaborative and democratic approach to solve societal problems. Furthermore, this approach creates spaces and opportunities for knowledge exchange and transfer between different stakeholders (Hansson & Polk, 2018).

Another aspect of knowledge co-production is 'social learning for interdependence', which refers to bridging the gap between an expert and the people who possess experiential knowledge (Bouwen & Taillieu, 2004). Undocumented knowledge possessed by the community members (about the history of canals, story of progressive pollution, tacit knowledge about canals) does not have the legitimacy accorded to published literature. According to Cash *et al.* (2005), for effective knowledge to take place and for knowledge to cross the boundary between organisations, it is important that negotiation or creation of knowledge is found to be credible, legitimate and salient by all the stakeholders. In this regard, salience describes the relevance and contextual accuracy of the information on which the collaborating stakeholders base their decisions. For example, the appropriate technology solution of DEWATS for the Municipal Colony portrays high salience. Credibility refers to the extent to which scientific outputs and outcomes are perceived as accurate, valid, robust and of high quality by decision makers, while legitimacy describes the extent to which the knowledge producers are considered to be unbiased and trustworthy by decision makers. IIT Bombay is one of the leading academic institutions in the country, and the participation of researchers from the institute provided credibility and legitimacy to the initiative.

Starting with the first winter school in 2017, IIT Bombay and the student participants of summer/winter schools created detailed knowledge about the sanitation practices in Alleppey including knowledge from the experiences and perspectives of non-scientific stakeholders like the residents who were surveyed during the summer/winter schools. This was then converted into different forms – reports, presentations, documentaries

---

[2] According to Census 2011, the literacy rate of Alleppey district is 96.24%.

and MOOCs. And different mechanisms were employed for disseminating the knowledge between different stakeholders.

Thus, the generated knowledge was distributed through a knowledge network which exhibited features identified by Luijendijk & Mejia-Velez (2005). First, the knowledge production exercises like summer/winter schools sought to create value by moving beyond information sharing so as to create new knowledge that translated into solution proposals that served as the basis for interventions. Secondly, the activities directly contributed to improving the capacity – through communication and knowledge exchange of the participants – volunteers of CANALPY, school students and interns from academic institutions and officials of line departments. Finally, these activities helped the CANALPY team members in networking with potential stakeholders and in devising strategies for engaging with them as well as other decision makers. These stakeholders became the key collaborators of various activities of the initiative in the field.

CANALPY aimed to leverage the potential for collaboration between different institutions and provided a platform for the same. Since it is difficult to foresee resources and needs for the initiative, collaboration is one of the ways to identify them (Beier *et al.*, 2017). Collaboration between different government agencies was achieved by inviting coordination and support from them. Coordinating with these agencies ensured strategic and nonlinear collaboration which is illustrated in Figure 4. Since the number of collaborating organisations was small, it was manageable and the shared jurisdiction also played a role in the collaboration concluding in local action. This particular characteristic is suggested as one of the contextual features that make a transdisciplinary project

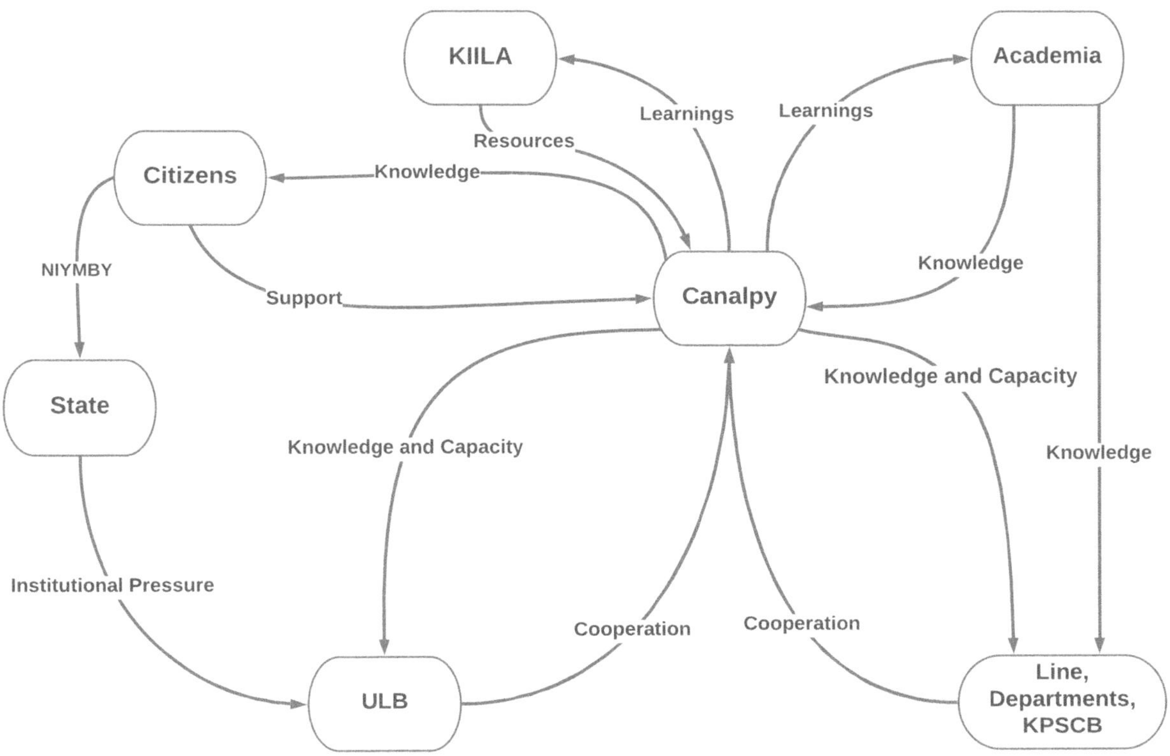

Fig. 4. | Relationships between collaborators in the network.

possible in a field setting (Margerum, 2008). Collaboration was beneficial to researchers involved in the initiative as well, as it enhanced their 'co-productive capacities' that are a combination of the scientific resources, cognitive, social and normative capacities to connect knowledge with action (Djenontin & Meadow, 2018). The team from IIT Bombay used the protocols and knowledge generated in Alleppey and leveraged the network of student volunteers to replicate the experiment in the neighbouring region of Kuttanad.

Another important factor that enables transdisciplinary local action projects is the mobilisation of resources. Since collaboration includes parties with various mandates, interests and expertise, it ensures that a complex problem is broken down into simpler questions and solutions are achieved at lower cost and without delays (Beier *et al.*, 2017). This is vital especially in the case of water and sanitation, which is often the responsibility of local governments that normally function with limited staff and financial resources (Crocker *et al.*, 2016). The KILA played an important role as the main funding source for the knowledge production activities and provided the financial and human resources. Furthermore, both formal networks and informal engagement of team members with the residents yielded necessary resources. For example, discussion through the formal network helped in identifying CSR funds for the construction of the DEWATS plant. Furthermore, informal conversations between the CANALPY team, interns and residents were instrumental in the canal clean-up campaign where local residents contributed by participating as volunteers as well as providing tools and food to the volunteers.

Factors at the individual level

At the individual level, leadership, agency and reflection play an important role in facilitating collaboration and knowledge co-production (Patterson *et al.*, 2013). Leadership leverages some aspects of human capital so that individual community members can mobilise resources and influence action by acting as change agents or leaders (Bass 1990 cited in Chaskin, 2001). Leadership includes both formal and informal leaders (Goodman *et al.*, 1998): formal political leadership in the form of support of the local Member of Legislative Assembly (MLA), the Finance Minister of Kerala and one of the strongest enablers of CANALPY. This stems from his conviction as an ex-academic as well as his activities on decentralised solid waste management in the town that is hailed as a model that won national and international acclaim including recognition from UNEP.[3] Similarly, the local project coordinator, an elected representative to the Municipal Council leader, could keep CANALPY in touch with the local government. He also leveraged his connections and networks to mobilise local volunteers for surveys and studies. The coordinator and main volunteer is also a dynamic activist who has attributed the success of the activities to their connection with the civil society organisation of KSSP. In the case of multiple collaborating institutions, it becomes important to recognise the power differences and have a neutral role as a facilitator rather than a hierarchical leadership setup (Bouwen & Taillieu, 2004). Thus, the local coordinators fulfilled the leadership duties of providing logistical support, process support as well as agenda-setting, framing and guiding the initiative by engaging in strategic negotiation with the collaborators. Support was provided by certain interested residents who with their agency served as norm setters. To explain this, it is important to explore the 'agency' factor.

Transdisciplinary research approaches recognise the importance of human agency in transformational projects attempting to bring about systemic changes (van Breda & Swilling, 2019). Agency is defined as the capacity of people to create, change and live according to their own meaning systems and to have the power and capacity to work towards the community's own destiny (Bhattacharyya, 2004). Individual agency plays an important role in enhancing community capacity (Chaskin, 2001) and comprises human capital and leadership – skills, knowledge and resources (Chaskin, 2001). In the case of CANALPY, in Municipal Colony, a few key individuals

---

[3] https://www.unep.org/news-and-stories/story/solid-approach-waste-how-5-cities-are-beating-pollution (accessed 3 November 2019).

were the main source of agency, who through their leadership, trust-building and willingness to learn new things, triggered learning and action among other residents. This agency was evident from residents of the Municipal Colony reporting violators who throw waste into the canals and from residents volunteering their time, labour and resources for cleaning the canal. Frequent engagement of the team and the researchers with the community through documentary screening, discussions and meetings also led to strengthening community members' ability to assert their agency.

Another factor that has functioned as an enabler at the individual/community level is reflexivity. According to Schwandt (1997, p. 135) (cited in Waghid (2002)), reflexivity is 'the process of critical self-reflection on one's biases, theoretical predispositions, preferences and so on'. Transdisciplinary knowledge is produced by engaging researchers reflexively and pushes universities to shift from being producers of disciplinary knowledge to creatively configure knowledge for tackling increasingly complex problems (Waghid, 2002). In the case of CANALPY, researchers from IIT Bombay and the CANALPY team were constantly motivated to engage in group reflections and discussions regarding the choices and decisions made in the initiative. Further local volunteers and interns have been engaged in multi-loop learning (van Breda & Swilling, 2019) wherein after gathering knowledge and skills through initial training sessions and summer schools, they could reiterate the learning process by delivering the next training session or workshops for students, community members and new interns. This constant engagement made them critically aware of the problems at hand and the consequence of the learning process. This also contributed towards reflexive learning. This commitment to reflexivity by the team and researchers also meant that inferences from the finding of one activity informed the next activity, thus leading to an iterative process of adapting research design and execution.

Barriers to the co-production activities

Practical barriers

Knowledge co-production and transdisciplinary research often face multiple practical and organisational barriers. Practical barriers include lack of time and resources (Polk, 2015). In this initiative, one of the challenges of the summer/winter schools was the lack of local participants. While the expectation from the knowledge production programmes was that young people trained from the summer/winter school would be a part of the awareness campaigns and studies, the proportion of local participants hailing from Alleppey was found to be much less. Many of the interns and volunteers also left after short stints constraining local youth representation. The political leadership in Alleppey expected the CANALPY team members to deliver quick results, and consequently, the team members had to devote a considerable amount of time and energy to managing the expectations. The clash of interests and differences in motivations between the stakeholders as found in literature was clearly observed in the context too. For example, knowledge exchange between researchers, local officials and bureaucrats is difficult when they employ different approaches and have a different vocabulary for framing the problem (Polk, 2015). Also, high expectations of public partners like political representatives may also clash with the production of results on the ground by the project team (Lang *et al.*, 2012). All these come in the way of practical barriers to implement the plans meticulously made through a long process of data collection, iterative analysis, reflection and finalisation.

Socio-political barriers

Varied political interests of the stakeholders can delay or impede consensual decision-making by stakeholders (Ansell & Gash, 2007). In the case of Alleppey Municipality, since the local and state governments belong to adversarial political fronts, it was difficult to develop cooperative alliances for implementation. Inferences from the FGD suggested changing dynamics of local politics. The interest of local government and their participation was uneven mainly due to external pressure developments in state and local politics, specifically the local elections which took place in 2020. Earlier, the local office bearers of the opposition party wanted the activities of CANALPY to succeed, as it was an initiative of their party leader who was the MLA from that constituency and the Finance Minister of the state. But with impending local body elections and fear of the ruling coalition using CANALPY for electoral gains, the initiative was abandoned by the opposition party. Thus, the same enabling political factors that earlier facilitated the initiative changed into constraints later. Furthermore, the heterogeneous nature of the community affected the participation of the community and in turn the knowledge co-production process. The FGDs conducted clarified that communities are fragmented along gender, caste and class lines. Social hierarchies and stratifications lead to conflicting political agendas, interests and power differentials which were observed as barriers in general (Polk, 2014).

Institutional factors

While the co-production process is a contextualised exercise, it is also important to note that the agencies collaborating on the initiative may not be designed to incorporate the demands of the project on their resources (Polk, 2015). The institutional culture and mindsets may not be suitable for a complex transdisciplinary project, as it entails institutional transaction costs – since various activities of the initiative were taking place outside the confines of their organisations, the stakeholders had to devote time, attention and other resources to the initiative. For academics involved in the initiative, time spent for travelling and attending meetings related to the activities of the initiative may lead to diversion of attention and time from other academic demands (Hoeft *et al.*, 2015). In CANALPY, the main academic partner IIT Bombay is not a proximate stakeholder impeding daily interactions. Due to limitations in motivations and institutional constraints, no local academic institution could take up the role, although attempts were made along those lines. Similarly, since the knowledge co-production and transfer involves multiple organisations, the process is prone to institutional inertia (Campbell, 2004). The local bureaucracy was fragmented along sectoral lines with no clear roles and responsibility definitions to involve in such an unconventional project. Likewise, the opportunity costs for participating stakeholders may be high and often not evident to their academic counterparts. Further logistical factors like lack of resources, financial and time constraints and geographical distance were also important issues faced during various activities of the initiative. These constraints shape the way activities are planned and may ultimately lead to ineffective engagement (Fritz & Binder, 2018).

Competence and knowledge gaps

Apart from practical and institutional constraints, competences of collaborating stakeholders play an important role in the participation and knowledge co-production process (Steelman *et al.*, 2015). Some of the local academic collaborators of CANALPY did not have the requisite skills and competencies to follow up on the various activities or integrate the various types of knowledge generated during the initiative. Academics and other collaborators in the project are trained in specific scientific disciplines and it rarely gives them the skills and competences needed to collaborate with diverse teams (Fritz & Binder, 2018). Many of the activities associated with the initiative could not be followed up due to lack of capacity and skills. This can be attributed to the fundamentals of the academic system which is perceived to be constraining and eager to slot students into disciplinary silos. Most of the disciplinary programmes are limited in providing training which fosters competencies needed for transdisciplinary and participatory research (Rozance *et al.*, 2020). Thus, this lack of skills and experiences may cause some academic researchers to find the whole process stressful and challenging (Cvitanovic *et al.*, 2019).

Apart from institutional factors, another important aspect that affects the transdisciplinary knowledge co-production process is the mindset of the collaborating stakeholders. While Hansson & Polk (2018) contend that high-level administrators with more power tend to be constrained in their thinking, CANALPY found some support from senior administrators of the local government and other agencies like the State Pollution Control Board who welcomed the new ideas and approaches in addressing the problem. However, some of the administrators were unfamiliar with updated knowledge systems, and the team encountered some resistance while trying to pitch alternate ideas to them. Furthermore, during the implementation of the DEWATS system, the team faced an inadvertent delay in bill payments due to hierarchies and red tape which prevented receptivity in organisations and deferred decision-making.

DISCUSSION

This study attempted to identify the drivers, motivations, barriers and challenges to the knowledge production process in the context of the CANALPY initiative in Kerala. One of the first drivers identified was the socio-political history of the state which was instrumental in steering the population towards democratisation, literacy and deepening of local governance. This is in line with findings from other scholars (O'Connor *et al.*, 2019) who posit that democratisation gives voice to citizens from a broad range of backgrounds thus empowering them to encourage decision makers to be open to all sources of knowledge. Furthermore, our findings also show that the process of decision-making is not only a function of knowledge but dependent on social learning. Experiential knowledge and expert knowledge interacted through collaborative processes to forge long-term relationships between the stakeholders and in building trust.

Our empirical analysis also reinforces findings from previous studies (Gerger Swartling *et al.*, 2019) that the assignment of an actor with leadership and collaborative skills for the task of facilitating and managing the process is central. This ensured support from the community and trust built with collaborators ensured continuity, especially the engagements with the local government. The findings also demonstrate that collaborating with multiple stakeholders has capacitated the local government and enhanced the co-productive capacities of the researchers. As in the case with other examples of knowledge co-production (Reyers *et al.*, 2015), the approach of creating platforms for collaboration has resulted in unearthing finer details of waste management practices which has led to the newer and contextual understanding that helped plan subsequent actions.

One of the limitations of the CANALPY initiative is the absence of any analysis of the impact of the activities on the local and other stakeholders. It is still an ongoing exercise, and an external evaluation is currently being explored by the team. This may provide suitable pointers to better design similar transdisciplinary initiatives to bring together academia, local community and local governments.

CONCLUSION

The study assessed CANALPY, a transdisciplinary initiative for knowledge co-production to address the capacity gaps in tackling the sanitation problem in a small town, particularly to rejuvenate its canals. Researchers involved in the initiative worked with local volunteers, citizens, government officials and the ULB for the production and transfer of knowledge to generate context-specific interventions. This study has sought to explore the knowledge co-production process and the various enabling drivers and barriers to the process. Examining the enablers through a multi-level framework helped identify factors that facilitated various activities of the initiative. The process of knowledge co-production is also fraught with risks and limitations mainly due to the socio-political, practical and institutional barriers. In summary, CANALPY has been instrumental in creating a space and platform for collaborations, dialogues and discussions on problems of sanitation, water quality and pollution. Crowdsourcing of data through large academic exercises like summer and winter schools helped generate a fine-grained analysis of sanitation problems at the town level. By converting the research into different types of knowledge products, CANALPY has ensured that the research findings got disseminated among a larger audience. These knowledge products have inspired new thinking and action, especially innovations like DEWATS in the Municipal Colony that points to the importance of sanitation planning to arrive at priorities for interventions, especially to achieve inclusive sanitation. The MOOCs designed for the elected representatives and officials of ULBs will help scale up the Alleppey experience to the entire state. Although the initiative attempted to bridge the capacity gaps in government agencies, it faced multiple barriers during the process. Another limitation of this study is the role of the authors in the initiative. By being actively involved in the initiative, there is the risk of our objectivity being compromised. However, we have attempted to periodically reflect on our roles as researchers critiquing the initiative as well as our role in the same. Many themes, which have a direct impact on the knowledge co-production and exchange such as the role of leadership, trust and partisan politics, also remain under-explored in this study. Future work has to analyse these along with evaluating the larger implications and outcomes of the initiative.

DATA AVAILABILITY STATEMENT

All relevant data are included in the paper or its Supplementary Information.

REFERENCES

Alappuzha Municipality & Government of Kerala (2014). *Master Plan for Alappuzha Town 2031.*

Ansell, C. & Gash, A. (2007). Collaborative governance in theory and practice. *Journal of Public Administration Research and Theory 18*(4), 543–571. https://doi.org/10.1093/jopart/mum032.

Bass, B. M. (1990). *Bass & Stogdill's Handbook of Leadership: Theory, Research, and Managerial Applications* (3rd. ed.). Free Press, New York.

Beier, P., Hansen, L. J., Helbrecht, L. & Behar, D. (2017). A how-to guide for coproduction of actionable science. *Conservation Letters 10*(3), 288–296. https://doi.org/10.1111/conl.12300.

Bhattacharyya, J. (2004). Theorizing community development. *Journal of Community Development Society 34*(2). https://doi.org/10.1080/15575330409490110

Bhosale, D. & Kumar, C. P. (2002). Simulation of seawater intrusion in Ernakulam coast. In: *Proc. of International Conference on Hydrology and Watershed Management,* (B. Venkateswara Rao, K. Ramamohan Reddy, C. Sarala & K. Raju eds.) Hyderabad, December 18–20, 2002, vol. II, pp. 390–399.

Bouwen, R. & Taillieu, T. (2004). Multi-party collaboration as social learning for interdependence: developing relational knowing for sustainable natural resource management. *Journal of Community and Applied Social Psychology 14*(3), 137–153. https://doi.org/10.1002/casp.777.

Brown, A., Dayal, A. & Rumbaitis Del Rio, C. (2012). From practice to theory: emerging lessons from Asia for building urban climate change resilience. *Environment and Urbanization 24*(2), 531–556. https://doi.org/10.1177/0956247812456490.

Bryman, A. (2012). *Social Research Methods*, 4th edn. Oxford University Press. https://doi.org/10.1192/bjp.112.483.211-a.

Campbell, J. L. (2004). *Institutional Change and Globalization.* Princeton University Press, Princeton, NJ.

Cash, D., Clark, W. C., Alcock, F., Dickson, N., Eckley, N. & Jager, J. (2005). Salience, credibility, legitimacy and boundaries: linking research, assessment and decision making. *SSRN Electronic Journal.* https://doi.org/10.2139/ssrn.372280.

Chaskin, R. J. (2001). Urban affairs review case studies from a comprehensive. *Urban Affairs Review 36*(3), 291–323. https://doi.org/10.1177/10780870122184876.

Chhajed-Picha, P. & Narayanan, N. C. (2021). Refining the shit flow diagram using the capacity-building approach–Method and demonstration in a south Indian town. *Journal of Environmental Management 294*, 112971.

Chong, J., Abeysuriya, K., Hidayat, L., Sulistio, H. & Willetts, J. (2016). Strengthening local governance arrangements for sanitation: case studies of small cities in Indonesia. *Aquatic Procedia 6*, 64–73. https://doi.org/10.1016/j.aqpro.2016.06.008.

Crocker, J., Shields, K. F., Venkataramanan, V., Saywell, D. & Bartram, J. (2016). Social science & medicine building capacity for water, sanitation, and hygiene programming: training evaluation theory applied to CLTS management training in Kenya. *Social Science & Medicine 166*, 66–76. https://doi.org/10.1016/j.socscimed.2016.08.008.

Croese, S. (2021). Co-producing urban knowledge in Angola and Mozambique: towards meeting SDG 11. *NPJ Urban Sustainability*, 1–10. https://doi.org/10.1038/s42949-020-00006-6.

Cvitanovic, C., Howden, M., Colvin, R. M., Norström, A., Meadow, A. M. & Addison, P. F. E. (2019). Maximising the benefits of participatory climate adaptation research by understanding and managing the associated challenges and risks. *Environmental Science and Policy 94*, 20–31. https://doi.org/10.1016/j.envsci.2018.12.028.

Djenontin, I. N. S. & Meadow, A. M. (2018). The art of co-production of knowledge in environmental sciences and management: lessons from international practice. *Environmental Management 61*(6), 885–903. https://doi.org/10.1007/s00267-018-1028-3.

Dreze, J. & Sen, A. (1997). *Indian Development: Selected Regional Perspectives.* Oxford University Press, Delhi.

Franke, R. W. (2019). *Life Is a Little Better: Redistribution as a Development Strategy in Nadur Village.* Routledge, Kerala.

Fritz, L. & Binder, C. R. (2018). Participation as relational space: a critical approach to analysing participation in sustainability research. *Sustainability 10*(8). https://doi.org/10.3390/su10082853

George, K. (1998). The Kerala model of development: a debate (Part 2). *Bulletin of Concerned Asian Scholars 30*(4), 35–40.

Gerger Swartling, Å., Tenggren, S., André, K. & Olsson, O. (2019). Joint knowledge production for improved climate services: insights from the Swedish forestry sector. *Environmental Policy and Governance 29*(2), 97–106. https://doi.org/10.1002/eet.1833.

GoK (2014). *Master Plan for Alappuzha Town 2031 (Draft).*

Goodman, R. M., Speers, M. A., Mcleroy, K., Fawcett, S., Kegler, M., Parker, E., Smith, S. R., Sterling, T. D. & Wallerstein, N. (1998). Identifying and defining the dimensions of community capacity to provide a basis for measurement. *Health Education and Behavior 25*, 258–278.

Hansson, S. & Polk, M. (2018). Assessing the impact of transdisciplinary research: the usefulness of relevance, credibility, and legitimacy for understanding the link between process and impact. *Research Evaluation 27*(2), 132–144. https://doi.org/10.1093/reseval/rvy004.

Heller, P. (2001). Moving the state: the politics of democratic decentralization in Kerala, South Africa, and Porto Alegre. *Politics and Society 29*(1), 131–163. https://doi.org/10.1177/0032329201029001006.

Heller, P., Harilal, K. N. & Chaudhuri, S. (2007). Building local democracy: evaluating the impact of decentralization in Kerala, India. *World Development 35*(4), 626–648. https://doi.org/10.1016/j.worlddev.2006.07.001.

Hoeft, T. J., Burke, W., Hopkins, S. E., Trinidad, S. B., James, R. D. & Boyer, B. B. (2015). Research: budgetary and other cost considerations. *Health Promotion Practice 15*(2), 263–270. https://doi.org/10.1177/1524839913485962.Building.

Hughes, S. (2013). Justice in urban climate change adaptation: criteria and application to Delhi. *Ecology and Society 18*(4). https://doi.org/10.5751/ES-05929-180448

Isaac, T. T. M. (1985). From caste consciousness to class consciousness Alleppey coir workers during inter-war period. *Economic and Political Weekly XX*(4), PE5–PE18.

Isaac, T. T. M. & Gopakumar, M. (2016). *Maarunna Manassukal, Maalinyamakalunna Theruvukal (Changing Hearts, Cleaning Refuse from the Streets).* Chintha Publishers, Trivandrum.

Jagannathan, K., Arnott, J. C., Wyborn, C., Klenk, N., Mach, K. J., Moss, R. H. & Sjostrom, K. D. (2020). Great expectations? Reconciling the aspiration, outcome, and possibility of co-production. *Current Opinion in Environmental Sustainability 42*, 22–29. https://doi.org/10.1016/j.cosust.2019.11.010.

Jahn, T., Bergmann, M. & Keil, F. (2012). Transdisciplinarity: between mainstreaming and marginalization. *Ecological Economics 79*, 1–10. https://doi.org/10.1016/j.ecolecon.2012.04.017.

Karlaganis, C. & Narayanan, N. C. (2014). *Governance Challenges in Linking Environmental Sustainability to Tourism: Where Is the Houseboat Industry in Kerala, India Headed? (Issue 2013/09).*

Lang, D. J., Wiek, A., Bergmann, M., Stauffacher, M., Martens, P., Moll, P., Swilling, M. & Thomas, C. J. (2012). Transdisciplinary research in sustainability science: practice, principles, and challenges. *Sustainability Science 7*(suppl. 1), 25–43. https://doi.org/10.1007/s11625-011-0149-x.

Luijendijk, J. & Mejia-Velez, D. (2005). Knowledge networks for capacity building: a tool for achieving the MDGs. In: *Workshop Proceedings on Design and Implementation of Capacity Development Strategies*, Beijing, China.

Margerum, R. D. (2008). A typology of collaboration efforts in environmental management. *Environmental Management 41*(4), 487–500. https://doi.org/10.1007/s00267-008-9067-9.

Narayanan, N. C. (2003). *Against the Grain: The Political Ecology of Land Use in a Kerala Region.* Shaker Publishing, Maastricht.

Narayanan, N. C., Rana, N. & Sridhar, A. (2017a). *Participatory Water and Sanitation Assessment Participatory Wastewater Assessment for Alappuzha Town*. Preliminary Report and Way Forward, pp. 1–62.

Narayanan, N. C., Rana, N., Sridhar, A., George, S., Menon, R. & Joseph, R. (2017b). *Preliminary Water and Sanitation Assessment for Alappuzha Town*. No. 1, CANALPY Working Paper.

Narayanan, N. C., Kumar, S., Rana, N., PIllai, S., Burt, Z. & Chhajed, P. (2018a). *Canal Rejuvenation in Alappuzha Town: A Multidisciplinary Study*.

Narayanan, N. C., Ray, I., Gopakumar, G. & Argade, P. (2018b). Towards sustainable urban sanitation: a capacity-building approach to wastewater mapping for small towns in India. *Journal of Water Sanitation and Hygiene for Development 8*(2), 227–237. https://doi.org/10.2166/washdev.2017.071.

O'Connor, R. A., Nel, J. L., Roux, D. J., Lim-Camacho, L., van Kerkhoff, L. & Leach, J. (2019). Principles for evaluating knowledge co-production in natural resource management: incorporating decision-maker values. *Journal of Environmental Management 249*, 109392. https://doi.org/10.1016/j.jenvman.2019.109392.

Patterson, J. J., Smith, C. & Bellamy, J. (2013). Understanding enabling capacities for managing the 'wicked problem' of nonpoint source water pollution in catchments: a conceptual framework. *Journal of Environmental Management 128*, 441–452. https://doi.org/10.1016/j.jenvman.2013.05.033.

Peters, G. B. (2015). *Advanced Introduction to Public Policy*. Edward Elgar Publishing, Cheltenham, UK.

Polk, M. (2014). Achieving the promise of transdisciplinarity: a critical exploration of the relationship between transdisciplinary research and societal problem solving. *Sustainability Science 9*(4), 439–451. https://doi.org/10.1007/s11625-014-0247-7.

Polk, M. (2015). Transdisciplinary co-production: designing and testing a transdisciplinary research framework for societal problem solving. *Futures 65*, 110–122. https://doi.org/10.1016/j.futures.2014.11.001.

Porio, E. (2011). Vulnerability, adaptation, and resilience to floods and climate change-related risks among marginal, riverine communities in Metro Manila. *Asian Journal of Social Science 39*(4), 425–445.

Quamar, R., Janipella, R., Jangam, C., Balwant, P., Veligeti, J., Chintalapudi, P. & Pujari, P. R. (2018). A case study of on-site sanitation system on the quality of groundwater in hard rock terrain of Coimbatore city. *Environment, Development and Sustainability 21*(4), 1–13. https://doi.org/10.1007/s10668-018-0116-0.

Reyers, B., Nel, J. L., O'Farrell, P. J., Sitas, N. & Nel, D. C. (2015). Navigating complexity through knowledge coproduction: mainstreaming ecosystem services into disaster risk reduction. *Proceedings of the National Academy of Sciences of the United States of America 112*(24), 7362–7368. https://doi.org/10.1073/pnas.1414374112.

Rosenqvist, T., Mitchell, C. & Willetts, J. (2016). A short history of how we think and talk about sanitation services and why it matters. *Journal of Water Sanitation and Hygiene for Development 6*(2), 298–312. https://doi.org/10.2166/washdev.2016.118.

Rozance, M. A., Krosby, M., Meadow, A. M., Snover, A., Ferguson, D. B. & Owen, G. (2020). Building capacity for societally engaged climate science by transforming science training. *Environmental Research Letters 15*, 12. https://doi.org/10.1088/1748-9326/abc27a.

Schwandt, T. A. (1997). *Qualitative Inquiry: A Dictionary of Terms*. Sage Publications, Inc, Thousand Oaks, CA.

Sharma, D., Singh, R. & Singh, R. (2014). Building urban climate resilience: learning from the ACCCRN experience in India. *International Journal of Urban Sustainable Development 6*(2), 133–153. https://doi.org/10.1080/19463138.2014.937720.

Shrum, W. & Ramanathaiyer, S. (2000). Knowledge, democratization, and sustainability: the Kerala 'model' of scientific capacity building. In: *Kerala: the Development Experience* (G. Parayil ed.). Zed Books, London, pp. 157–177.

Smith, L. E. D. & Porter, K. S. (2010). Management of catchments for the protection of water resources: drawing on the New York City watershed experience. *Regional Environmental Change 10*(4), 311–326. https://doi.org/10.1007/s10113-009-0102-z.

Steelman, T., Nichols, E. G., James, A., Bradford, L., Ebersöhn, L., Scherman, V., Omidire, F., Bunn, D. N., Twine, W. & McHale, M. R. (2015). Practicing the science of sustainability: the challenges of transdisciplinarity in a developing world context. *Sustainability Science 10*(4), 581–599. https://doi.org/10.1007/s11625-015-0334-4.

Subramanyam, N. (2020). A small improvement: small cities lag in expanding household water coverage across urban India. *Water Policy 22*(3), 468–482. https://doi.org/10.2166/wp.2020.116.

Talwar, S., Wiek, A. & Robinson, J. (2011). User engagement in sustainability research. *Science and Public Policy 38*(5), 379–390. https://doi.org/10.3152/030234211X12960315267615.

Tutusasus, M. & Schwartz, K. (2018). Water services in small towns in developing countries: at the tail end of development. *Water Policy 20*(S1), 1–11. https://doi.org/10.2166/wp.2018.001.

van Breda, J. & Swilling, M. (2019). The guiding logics and principles for designing emergent transdisciplinary research processes: learning experiences and reflections from a transdisciplinary urban case study in Enkanini informal settlement, South Africa. *Sustainability Science 14*(3), 823–841. https://doi.org/10.1007/s11625-018-0606-x.
Waghid, Y. (2002). *Knowledge Production and Higher Education Transformation in South Africa: Towards Reflexivity in University Teaching, Research and Community Service. Higher Education 43*(4), 457–488.
Wiek, A., Talwar, S., O'Shea, M. & Robinson, J. (2014). Toward a methodological scheme for capturing societal effects of participatory sustainability research. *Research Evaluation 23*(2), 117–132. https://doi.org/10.1093/reseval/rvt031.

First received 21 April 2021; accepted in revised form 30 September 2021. Available online 18 October 2021

© 2022 The Authors From Knowledge and Capacity Development to an Implementation Science: doi: 10.2166/wp.2021.263

Exploratory assessment of challenges and issues with private water operators in rural water supply and service delivery: a case study of the Karamoja region, Uganda

Benbella Dektar[a,*], Scott McConnell[b] and Allan Kasekende[b]

[a] World Bank's Somalia Urban Resilience Project, Phase II (SURP II), Department of Public Works, Reconstruction & Housing, Mogadishu, Somalia
[b] MK Dev Consults Limited, Department of WASH Business Development, Nairobi, Kenya
*Corresponding author. E-mail: dektarb@gmail.com

ABSTRACT

This exploratory assessment was conducted in the Karamoja region of North Eastern Uganda. The aim was to analyse the contextual shortfalls that undermine the effectiveness and sustainability of private operators (PO) in water supply. POs had limited capacity to manage affairs of the water supply systems (WSS). Specifically, their skills were lacking to manage the electromechanical components of the systems, meter reading, and they exhibited poor customer care skills. Intermittent power supplies rendered the WSS unreliable. Solar systems were reported to have erratic frequencies and output which also ultimately resulted in inconsistent water supply. However, the POs hardly pursued corrective measures. The weak enforcement of regulatory policies propagated non-uniformity in tariff implementation approaches across districts contributing to lower willingness to pay. The POs' focus on profits presents a unique challenge as they attempt to minimise energy expenses in a bid to widen profit margins. We conclude that the limited technical and managerial capacity of POs in running water supply systems is a critical cause of unsatisfactory service delivery to water users. Finally, the weak water governance and regulatory policy enforcement and the non-adherence to water tariff plans not only affect PO economies of scale but also hamper efficient service delivery.

Key words: Capacity, Poverty, Private operators, Sustainability, Uganda, Water user

HIGHLIGHTS

- Institutional capacity weakness exists in sectoral coordination where it is least expected to manifest at policy formulation level which ultimately impedes the oversight function of the water authority.
- Inconsistent data use to strengthen data-driven iterative adaptation to service delivery is a major factor that inhibits evidence-based decision making.
- Weak regulatory policy enforcement mechanisms stymie water governance leading to inconsistent water tariffs, and intermittent water supply, and worse still system breakdown.

GRAPHICAL ABSTRACT

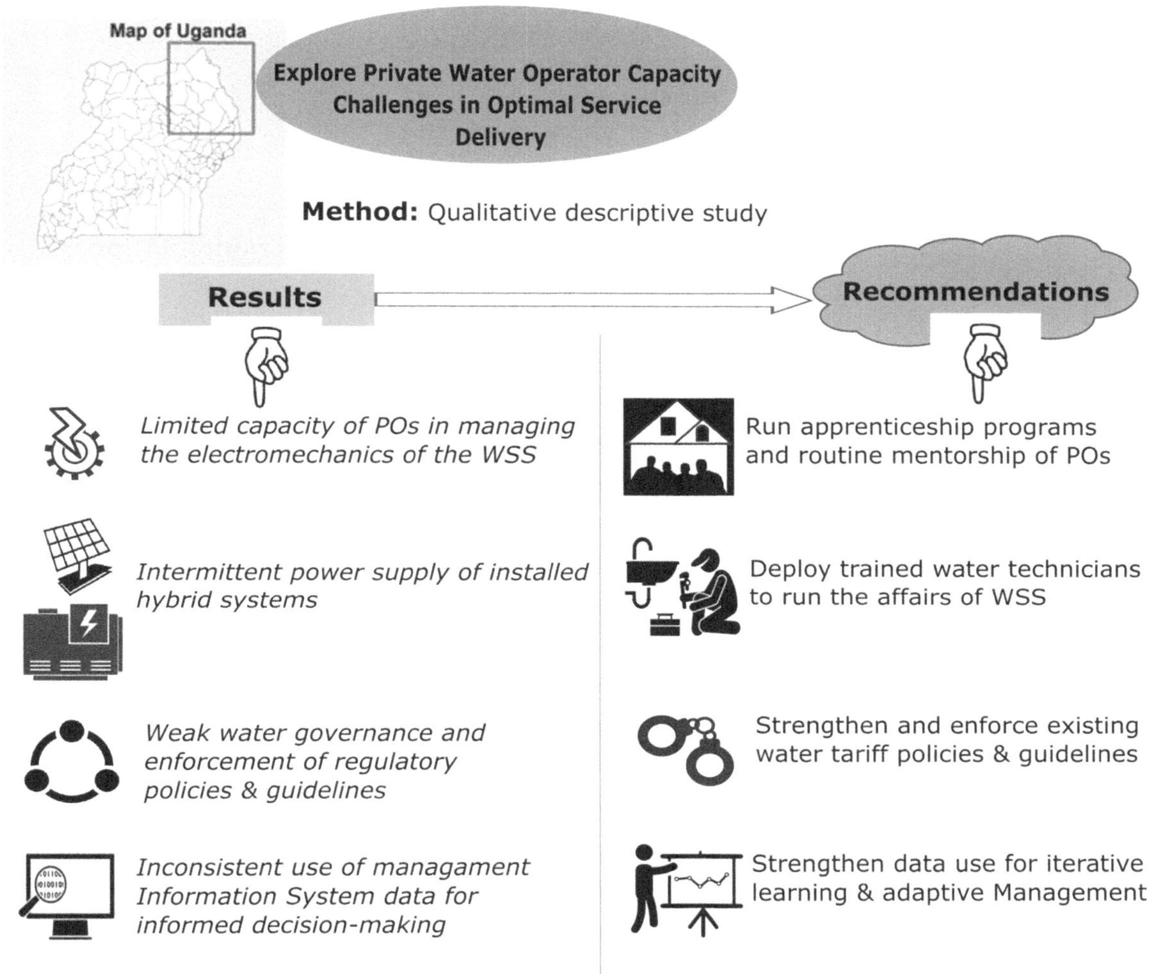

Figure 1 offers an at-a-glance visualisation of the exploratory assessment highlighting the method used, main results, and actionable recommendations. Representative symbols/icons have been used to reinforce the brief descriptions under the results and recommendations.

INTRODUCTION

Located in North Eastern Uganda, Karamoja is a semi-arid region characterized by erratic precipitation patterns and long drought spells. It is known to have the lowest groundwater potential due to low water tables. The average drilling depth and static water levels in the region are 79.3 m and 28.5 m respectively (DWRM & JICA, 2011). Karamoja is one of the underprivileged regions that has for decades been underserved in water and sanitation services by the Government of Uganda. Moreover, the region suffers from the worst poverty indicators in the country, at 74% (UBOS, 2018), a history of armed conflict and livestock rustling, poor road infrastructure,

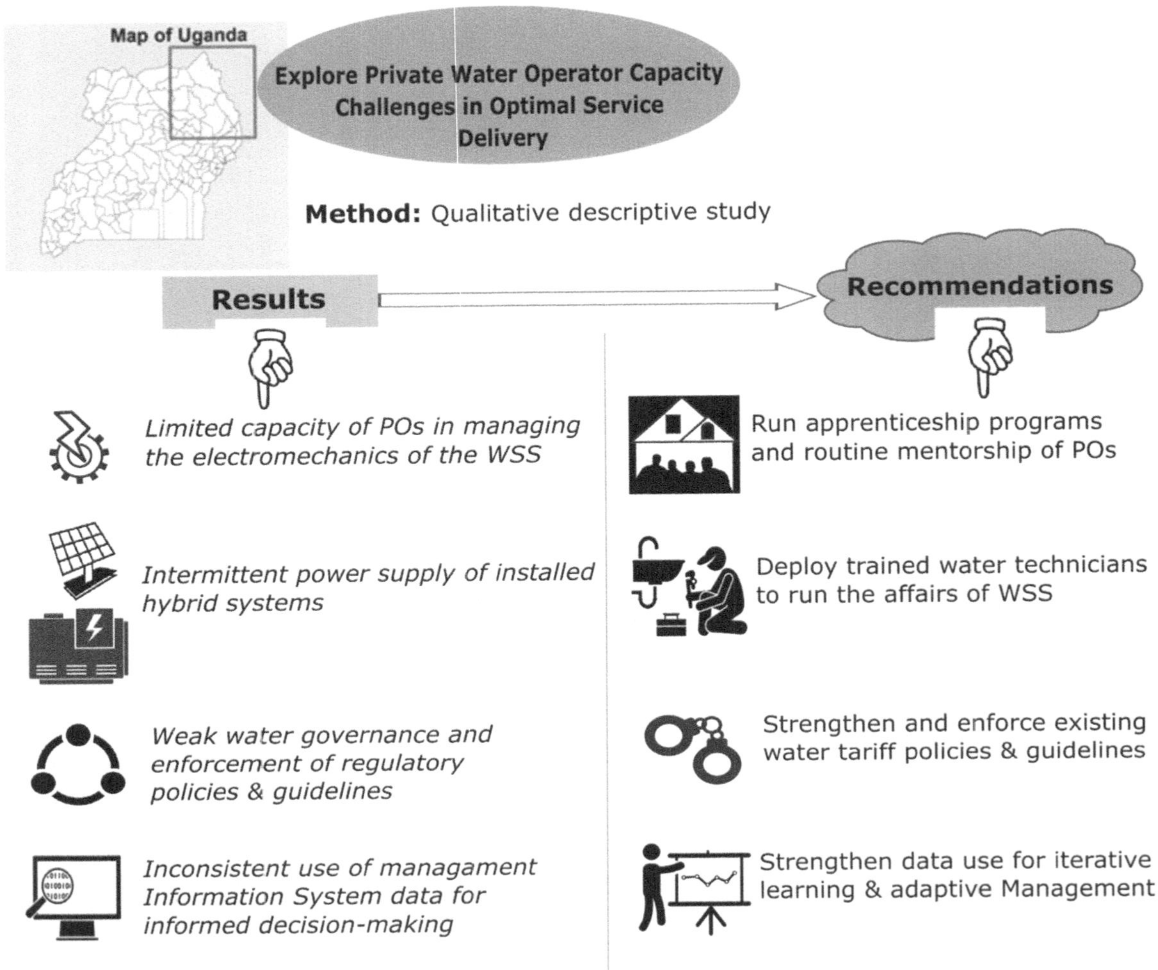

Fig. 1. | Graphical at-a-glance visualization of the method, results and recommendations.

lack of connection to the national power grid, and pastoralist lifestyles. These problems exacerbate a myriad of challenges that private water operators (POs) must address when setting water tariff plans and delivering the services in the area. Despite her wealth in mineral and natural resources, Karamoja continues to receive minimal attention from the government in terms of infrastructural development and social service delivery. According to the Ministry of Water and Environment (MWE), the region ranked lowest in sanitation and hygiene indicators albeit did comparatively well in water access. Specific water coverages for the districts of Moroto, Kotido, and Kaabong where the assessment was conducted stood at 77, 78, and 84%, respectively (Ministry of Water and Environment, 2017). These figures exceeded the national average safe water coverage in rural areas which was estimated at 70%. On the other hand, the average functionality for rural water supplies across Uganda was 85% whereas Kaabong and Moroto districts registered a functionality ranging between 76% and 84.9%, while Kotido fell below this range (Ministry of Water and Environment, 2018). We noted that the functionality of water supply systems (WSS) is largely dependent on the commitment and performance of the POs while recognizing that the Water Boards and consumers, too, have their role to play. Water boards are special entities that take responsibility and oversight of rural water supply and sanitation services. These figures align with for instance Kleemeier (2010) who reported that the continued functioning of rural water supplies in Africa remains problematic, despite much progress over the past three decades. In addition to continuing access expansion, the high breakdown rate due to lack of maintenance is threatening the sustainability of what has already been achieved (Foster & Briceno-Garmendia, 2010).

Ndaw (2016) attributes this failure of the WSS to the inability of local governments to provide WSS services after decentralization of water service delivery from central to local governments, among other factors. Ndaw's study adds that Local Governments often lack the necessary skills and capacity to manage water supply and sanitation services, which creates the case for private sector involvement with the aim to bridge the capacity gap. However, whereas POs are, in principle, able to overcome some of the shortfalls, they, too, are faced with other contextual challenges. For example, poor follow-up is given to the District Implementation Manual of MWE which highlights that key operation and maintenance (O&M) issues should be discussed with communities from the early stages of planning the water source improvement (MWE, 2013). That way, critical issues could be identified from the outset and addressed in time. Moreover, doing so is advantageous in gaining client confidence and consistency in tariff payments.

A study on service failures of rural water supply systems in Ethiopia revealed that lack of uniformity of implementation approaches, and institutional and organizational incapability of the local government aggravate the service failures (Behailu *et al.*, 2017). In Kenya, a study by Spaling *et al.* (2014) suggested that community resistance to compliance with water sector reforms, including those intended to benefit community-managed projects, negatively affect the sustainability of water supply projects. Whereas this study identifies this challenge, it is our view that, if good follow up mechanisms by the water authority and adequate Water Safety Plans (WSPs) are put in place, community management through existing local structures and scheme operators can be effective in the long run.

In Tanzania, the low quality of water services, the lack of sustainability of constructed infrastructure, the difficulties in targeting the poor, and the inadequate internal information systems are main weaknesses that continue undermining strategies for poverty eradication (Jimenez & Perez-Foguet, 2010). Finally, Kanyesigye *et al.* (2019) stressed the need for systematic documentation of experiences to demonstrate the benefits of WSP and narrate success stories as well as challenges encountered during implementation. This is essential for initiating the relevant improvements at the policy level notably concerning institutional arrangements, legal requirements and enforcement mechanisms.

Hence, POs are faced with a wide array of challenges which range from capacity gaps, complacency from clients, and infrastructural problems presented by the contexts within which they run the WSS, while at the same time having to strike a balance between effectiveness in service delivery and profit margins that are expected to emerge from the water tariffs.

On the above account, while determining tariffs, POs often hardly consider the financial constraints consumers face. Economic Consulting Associates (2015) reported that POs propose tariffs to be charged. The MWE approves the tariffs and signs the performance contracts with Water Supply and Sanitation Boards and the National Water and Sewerage Corporation (NWSC). One of the current problems is that delays in tariff revision and setting are common because of delays in the approval process. These delays indicate potential pitfalls in service delivery even when the POs are committed to offering services to more people. Ndaw (2016) reported that POs expressed a desire to expand their business to neighbouring unserved areas with evident demand, and recommended to allow for these types of expansions or the clustering of small towns.

Finally, it is worth noting that documentation of the operations of the private and scheme operators in the Karamoja region has been minimal, leading to a paucity of information on the topic at hand in this complex region. Moreover, the activities of POs in the rural areas outside of major towns and cities have not been studied well.

Therefore, we undertook this study to examine these issues in a structured way with the aim of documenting the challenges and pathways associated with enhancing effectivity in the operations of WSS operators but also contribute to the existing body of knowledge on POs running WSS.

Specifically, we sought to qualitatively explore the capacity enhancement challenges and pathways for POs to enhance effectivity in service delivery in three districts of the Karamoja region. We aimed at identifying and addressing the shortfalls that undermine the effectiveness of POs as key actors in water supply and sustainability.

METHODS

This research is based on case studies and was conducted between February and May 2019 in four water supply schemes located in the districts of Moroto, Kaabong, and Kotido in North Eastern Uganda (Figure 2).

Purposive sampling was used to select staff or members from the POs, Water Supply and Sanitation Boards (WSSB), and the Karamoja Umbrella for Water and Sanitation (KUWS), a coordination body, for interviews. Simple random sampling was used to identify water users (beneficiaries) from the lists of household heads

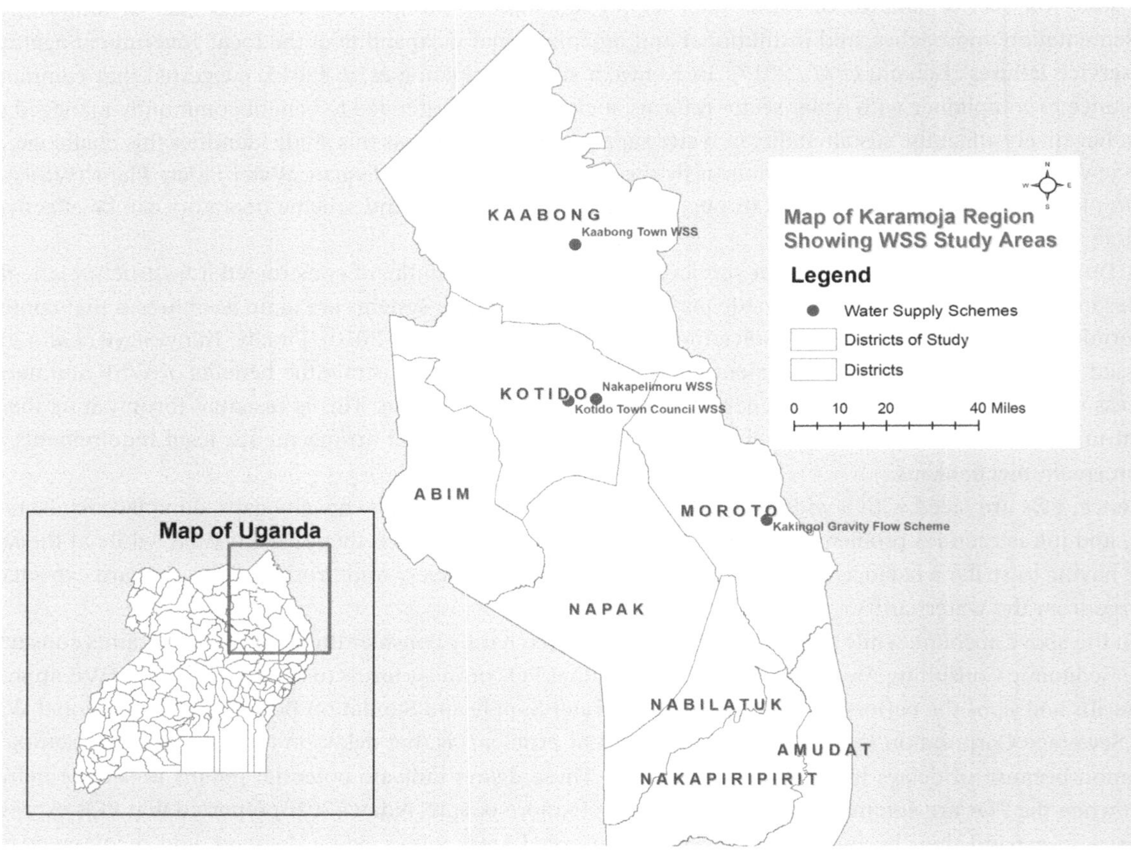

Fig. 2. | Location of the water supply schemes investigated.

maintained by local authorities for focus group discussions. We adopted a qualitative study design that mapped perceptions, experiences, and opinions of water users, WSSBs and POs.

To ensure meaningful triangulation, data was collected using a blend of qualitative data collection techniques and instruments (Table 1). Key Informant Interviews were used to obtain data from individuals with detailed knowledge of operations and processes involved in managing WSS. On the other hand, focus group discussions facilitated the extraction of detailed information from water users regarding, specifically, tariff and billing systems, customer satisfaction, and consistency in water supply. In-depth interviews ensured that the study obtained a detailed account of the experiences and practices of running WSS by private operators.

Observations were used to ascertain compliance with the general standard operational procedures (SOP) set out by the Ministry of Water and Environment which is the water authority.

The thematic analysis technique was used to manually analyse emergent latent and manifest content/data from interviews and discussions to uncover meanings in data and reveal hidden complexities. Thematic analysis is a type of qualitative analysis used to analyse classifications and present themes (patterns) that relate to the data. It illustrates the data in great detail and deals with diverse subjects via interpretations (Boyatzis, 1998).

Namey *et al.* (2008) stated that the 'thematic moves beyond counting explicit words or phrases and focuses on identifying and describing both implicit and explicit ideas.' In some instances, it may be better to use manual analysis rather than computer-based methods such as Nvivo (Alhojailan, 2012). Welsh (2002) argued that software might not prove as helpful as one may expect in terms of addressing issues of validity and reliability in thematic ideas that emerge during the data analysis process due to the fluid and creative way in which these themes emerge. Where speech was quoted, pseudonyms were used to protect study participants' identities. Direct quotations provide supportive meaning to the data's interpretation for some statements (Patton, 2002).

We obtained information on each of the water supply schemes visited (Table 2). Estimates of the population were obtained from the detailed parish files published by the Bureau of Statistics based on the 2014 National Population and Housing Census (UBOS, 2016). WSS information was obtained from unpublished technical drawings, schematics and other records availed by the District Water Offices of the Local Governments. Whereas obtaining copies of most of these records was prohibited, we were authorized to view them and take notes. Additional data was obtained from interviews and observations.

RESULTS AND DISCUSSION

Results could be categorized into three specific challenges.

Challenges related to private operators' capacity

Limited capacity to manage the water supply systems

The results showed that POs had limited capacity to manage affairs of the WSSs. A review of the available PO records for quarters II and III which spanned Oct–Dec 2018 and Jan–Mar 2019 respectively indicated that

Table 1 | Data collection methods, instruments and respondents.

#	Method	Respondents	Target	%
1.	In-depth Interviews (IDI)	POs	8	100%
2.	Key Informant Interviews (KII)	WSSB, KUWS	20	90%
3.	Focus Group Discussion (FGD)	Water users (clients)	08	88%
4.	Observation	POs, and water users	40	100%

Table 2 | Technical details of the water supply systems.

#	Facility Name	Type	Location [Prdn. Cap (m^3)]	Est. Popn served (UBOS, 2016)	Tank Volumes		# of PSP (fxn)	Private Operator	Power Source	Water treatment syst.
					Sedimentation	Reservoir				
1.	Kakingol GFS Commissioned: Oct 2009; ~8 km transmission pipeline	GFS	Moroto district 14 $m^3 day^{-1}$	3,250 (7 villages)	6.5 m^3	50 m^3 (Concrete)	13 (11)	Rotiyo Ltd 10 service connections	None (Gravitational flow)	HTH Chl. syst.
2.	Kotido Town Council WSS Commissioned: 1970s Comprises 3 production wells ~19 km trans. pipeline	WSS (M)	Kotido district 28 $m^3 day^{-1}$	13,838 (14 villages)	N/A	2 tanks (72 m^3, 100 m^3) HDPE	78 (78)	M/S Didi Ltd (NWSC) 494 service connections	Hybrid system [Solar PV arrays +45KVA Generator]	HTH Chl. syst.
3.	Nakapelimoru WSS Commissioned: 2009 Comprises 2 production wells ~10 km trans. p/line	WSS (M)	Kotido district 10.5 $m^3 hr^{-1}$	3,000 (11 villages)	N/A	4 tanks (10 m^3 each) HDPE	7 (7)	Local Scheme Operator (KUWS) 10 connections	Hybrid system	HTH Chl. syst.
4.	Kaabong Town WSS Commissioned: 2012 Comprises 1 Prdn well ~4.5 km p/line	WSS (M)	Kaabong district 21.25 $m^3 hr^{-1}$	11,559 (6 wards)	N/A	150 m^3 HDPE	7 (5)	M/S Trandint Ltd 263 connections	Hybrid system	Salt electrolysis (with Dosatron)

Legend: **fxn**- Functional; **GFS**, Gravity flow scheme; **HDPE**, High-density polyethylene; **HTH Chl. Syst**, High-Test Hypochlorite chlorination system; **KUWS**, Karamoja umbrella for water and sanitation; **KVA**, Kilo-volt amperes; **M**, Motorised; **N/A**, Not Applicable; **NWSC**, National water and sewerage corporation; **Prdn. Cap**, Production capacity (yield); **PSP**, Public stand post; **PV**, Photovoltaic; **RGC**, Rural growth centre; **WSS**, Water supply system.

operations were less than desirable as low scores were achieved on all parameters assessed. Scores were particularly low on the operations and maintenance (O&M) plans, the adequate and regular financial record keeping, and the frequency of meetings with the Water Supply and Sanitation Boards and the users. A lack of routine meetings with stakeholders – including water users – contravenes the District Implementation Manual (MWE, 2013), which advises that O&M issues should be discussed with communities from the early stages of planning the water source improvement.

Deployment of unskilled and inexperienced personnel by POs

The results highlighted notable challenges as follows: Limited and inadequate capacity in managing the electro-mechanical components of the WSS, and lack of basic skills in meter readings, minor repairs and customer care skills. About 80% of the technicians deployed by the POs lacked formal education and a basic skillset necessary to effectively run the WSS on behalf of their employers.

In Kaabong, Jimmy, a technician attending to the WSS reported that *'I have not been to a technical or vocational training institute, so, I am not educated but was trained for four days on how to run the system…'* These remarks indicate that there are inadequate trainings and mentorship offered to the technicians who manage the day-to-day affairs of WSS. This affirms Ndaw (2016)'s finding that local structures often lack the skills and capacity to manage water supplies and sanitation services.

Our findings further revealed that intermittent water supply resulted in conflicts that ultimately triggered events that led to vandalism of pump houses and public stand posts in some locations, especially in Kotido.

In Kotido district, Ms Jennifer (a client) said '…*The company managing our water doesn't have qualified people. The system is always breaking down and they take long to rectify the problems…*' Intermittent supply is often associated with the functionality of the WSS in terms of O&M practices and breakdown is inevitable if O&M routines are undermined either due to lack of technical capacity or negligence.

Challenges related to power supply

Erratic power supply renders the WSS unreliable

As indicated in Table 2, power is provided by diesel generators and solar panels. The solar-panel systems were reported to have erratic frequency and output during the day which resulted in intermittent water supply. The operators also run diesel generators; thus, high diesel prices pose great concerns on the reliability of water supply, especially in rainy seasons when the diesel trucks have difficulty using the road network. The power supply-related challenges are further exacerbated by the poor road infrastructure in the Karamoja region.

The fact that POs are seemingly unable to address these challenges is in fact primarily a problem of capacity given that they are unable to devise innovative mechanisms to counter these shortfalls.

Mr Robert, a member of the Water Board of Nakapelimoru stated that '…*sometimes power supply issues are beyond the scheme operator, solar power is dependent on sunlight which is minimal during the wet spell.*' And '*Our murram roads are very bad during rainy seasons. Trucks find it difficult to deliver fuel across Karamoja region, so, you can't blame the operator for that… it's beyond his control*'. Mr Lokiru, a water technician employed by one of the companies said: *'I wouldn't have a problem running the generator for extra hours to fill the reservoir tanks to capacity, unfortunately, I cannot do so because I occasionally run out of fuel yet the trucks take days to arrive due to the bad roads they have to go through.'*

This implies that intermittent water supply is inevitably aggravated by inconsistent fuel delivery due to bad roads, high prices notwithstanding. Thus, monetary gains for the PO cannot be guaranteed. However, these findings point towards inadequate preparedness in the operation and maintenance approaches. It was observed that

POs hardly stored sufficient fuel to run their systems for at least two weeks in the event of fuel shortage or limited supply.

Challenges related to water governance, regulatory policy, and clients

Lack of consultations/engagements between POs, WSSB, and the users

The above weaknesses led to deterioration in service delivery and resulted in disagreements between the PO and the water users especially on water tariff rates and general operation of the schemes. In the event, no opportunity or regular platform was created to discuss challenges, knowledge and capacity gaps, and identify remedial measures. Moreover, weak information hampered evidence-based decision making. For example, during the assessment, it was observed to be common for POs not to know the estimated population they serve. This also shows that there was hardly any consultation, nor interaction with the existing Utility Performance Monitoring & Information System (UPMIS) for piped water supply in Uganda. This online system was developed by the Directorate of Water Development of MWE to offer operational and performance information for WSS. Whereas this was a weakness and challenge at the WSS level, we also noted that the UPMIS was not regularly updated with recent data, hence rendering it an unreliable database for accurate real-time and valid data. This gap was one of the pointers to the capacity issues at the Directorate which inadvertently affects the operations and service delivery of POs.

Limited revenues due to inconsistent adherence to tariff payments

POs are compelled to run diesel generators for only approximately three hours a day, to minimize operational expenses which result in inadequate production and dry pipes within the network.

Limited production implies less revenue for the PO making the venture less attractive and compromising viability hence risking the collapse of the venture.

From Kaabong district, Mr Lopel, a private operator, affirmed that '*...as a private operator, I experience persistent defaults in water tariff payments by users, yet I am responsible for the technical and commercial operations of the scheme*'. He adds that '*...because of inconsistent tariff payments, I am unable to generate sufficient funds to effect expansions and run operations effectively.*'

Our study revealed that the inconsistent payment culture among the users which is complicated by wide spread poverty, among other domestic challenges, and a poor fee collection management strategy of the PO, pose challenges to both users and operators alike. Our study noted that tariff decisions and approval processes are bureaucratic. Even when decisions are made on tariffs, they hardly reach the consumers on time. This finding agrees with ECA (2015) that reported that POs hardly take into account the financial difficulties consumers face while determining tariffs. The report adds that one of the current problems is that delays in revising tariffs are common because of delays in the approval process. Clearly, the setting and implementation of realistic and effective tariff strategies require experience and capacity with the POs and the WSSBs, which seem seriously lacking.

Weak enforcement of regulatory policies by water authorities

The study observed weak enforcement practices by the WSSB, KUWS, and the District Water Offices of the Local Governments. Even where water tariff rates had been agreed upon, in some locations, we observed altercations between the POs and users regarding amounts to be paid. We also noted inconsistencies in tariff implementation approaches in Kotido and Kaabong which involved obtaining water on loan, payment in kind with food such as flour and other small household items. The loan arrangement is risky given that consumers are bound to default thereby leading to losses and limiting the achievement of economies of scale for the scheme operators.

Therefore, as concluded by Kanyesigye *et al.* (2019), initiating the relevant moves at the policy level such as enforcement mechanisms can be instrumental in improving service delivery. Weak enforcement often leads to consumer resistance to reforms and policies which affect long-term sustainability as also found by Spaling *et al.* (2014) who concluded that community resistance to compliance with water sector reforms, including those intended to benefit community-managed projects, affect the sustainability of water supply projects.

Similarly, Behailu *et al.* (2017) found that lack of uniformity in implementation approaches, and institutional and organizational incapability of the local government, aggravate the service failures. This implies that weak and inconsistent approaches in WSS management and governance coupled with weak policy enforcement jeopardise effective and efficient water service delivery.

CONCLUSIONS

Limited technical capacity of POs in running water supply systems is a central cause of unsatisfactory service delivery and is observed in the technical management, the unreliable power supply, the setting of credible tariffs and the collection thereof, the communication with the water users, and the collection of data for the management information system. Similarly, an important role exists for national authorities such as the NWSC to strengthen capacities of POs in key thematic areas such as business planning, O&M of WSS, water safety planning, asset management, financial management, etc. The capacity of POs to manage WSSs (including the technical capacity of staff) needs to be assessed prior to contract award. Such assessment at the contract award combined with regular apprenticeship of water technicians employed by POs will enhance the sustainability of WSS and ensure better service delivery to beneficiaries.

The second set of institutional capacity weakness exists in sectoral coordination. It is logical that POs attempt to generate enough profit, and POs are particularly enticed to minimize energy expenses. Therefore, national authorities such as MWE and NWSC need to strengthen their oversight and mentorship of POs and WSSBs on commercial operations including for staff involved in daily operations. Also, key policymakers from other sectors need to be approached to help address infrastructural challenges to water service delivery, notably the connection to the national power grid and the improvement of the road networks. The better coordination of the sectors is a critical condition to ensure more profitable ventures in undertakings of WSS by private and scheme operators.

Lack of routine consultations and engagements between the WSSBs and POs partly hampered service delivery. This missing link not only leads to deterioration in service delivery but also promotes distrust and lack of transparency between stakeholders. Therefore, regular deliberations are necessary as this strengthens and maintains relations ultimately leading to satisfactory service rendered to clients. In this vein, weak regulatory policy enforcement and the non-adherence to water tariff plans affect PO cash-flows.

DATA AVAILABILITY STATEMENT

All relevant data are included in the paper or its Supplementary Information.

REFERENCES

Alhojailan, M. I. (2012). Thematic analysis: a critical review of its process and evaluation. *West East Journal of Social Sciences 1*(1), 39–45.

Behailu, B. M., Jarmo, J. H. & Katko, T. S. (2017). Service failures of rural water supply systems in Ethiopia and their policy implications. *Public Works Management & Policy* 22(2), 179–196.

Boyatzis, R. E. (1998). *Transforming Qualitative Information: Thematic Analysis and Code Development*. SAGE Publications, California.

Directorate of Water Resources Management (DWRM), and the Japan International Cooperation Agency (JICA) (2011). *The Development Study on Water Resources Development and Management for Lake Kyoga Basin in Uganda. Report.*

Economic Consulting Associates (ECA) (2015). *Uganda Country Report in Strengthening Public Institutions in Engaging and Regulating Domestic Private Sector for the Provision of Water and Sanitation Services in Rural Growth Areas and Small Towns: Best Practice Report*. ECA, London.

Foster, V. & Briceno-Garmendia, C. (2010). *Africa's Infrastructure: A Time for Transformation: A Time for Transformation. Africa Development Forum*. World Bank.

Jimenez, A. & Perez-Foguet, A. (2010). Challenges for water governance in rural water supply: lessons learned from Tanzania. *International Journal of Water Resources Development 26*(2), 235–248.

Kanyesigye, C., Marks, S., Nakanjako, J., Kansiime, F. & Ferrero, G. (2019). Status of water safety plan development and implementation in Uganda. *International Journal of Environmental Research and Public Health 16*(21), 4096.

Kleemeier, E. L. (2010). *Private Operators and Rural Water Supplies: A Desk Review of Experience.* Water papers: World Bank, Washington, DC.

Ministry of Water and Environment (MWE) (2013). *District Implementation Manual.*

Ministry of Water and Environment (MWE) (2017). *Uganda Water and Environment Sector Performance Report 2017.*

Ministry of Water and Environment (MWE) (2018). *Uganda Water and Environment Sector Performance Report 2018.*

Namey, E., Guest, G., Thairu, L. & Johnson, L. (2008). Data reduction techniques for large qualitative data sets. *Handbook for Team-Based Qualitative Research.* In: (G. Guest & K. M. MacQueen, eds). Rowman Altamira Press, Maryland.

Ndaw, M. F. (2016). *Private Sector Provision of Water Supply and Sanitation Services in Rural Areas and Small Towns: The Role of the Public Sector.* Water and Sanitation Program Guidance Note: World Bank, Washington, DC.

Patton, M. Q. (2002). *Qualitative Research and Evaluation Methods*, 3rd edn. SAGE, California.

Spaling, H., Brouwer, G. & Njoka, J. (2014). Factors affecting the sustainability of a community water supply project in Kenya. *Development in Practice 24*(7), 797–811.

Uganda Bureau of Statistics, (UBOS) (2016). *The National Population and Housing Census 2014. Main Report.* Available at: https://www.ubos.org/wp-content/uploads/statistics/SAP_Northern.xlsx (accessed 19 April 2020)

Uganda Bureau of Statistics (UBOS) (2018). *Mapping the Spatial Distribution of Poor Households, Technical Report.*

Welsh, E. (2002). Dealing with data: using NVivo in the qualitative data analysis process. *Forum: Qualitative Social Research* 3, 2.

First received 11 December 2020; accepted in revised form 11 July 2021. Available online 28 July 2021

IWA Publishing's authorised EU representative for General Product Safety Regulations is Diane D'Arras, 15 rue Duret, 75116 Paris, France, e-mail: safety@iwap.co.uk.

Printed and bound by CPI Group (UK) Ltd, Croydon, CR0 4YY
12/05/2026
02108863-0020